Namaqualand in Flower

Sima Eliovson

NAMAQUALAND IN FLOWER

Macmillan · Johannesburg

First published in 1972 by
MACMILLAN SOUTH AFRICA (PUBLISHERS) (PTY) LTD
Johannesburg
Associated companies in London, Toronto,
New York, Dublin, Melbourne and Madras

SBN 0 86954 001 7

Photography and jacket painting by
Sima Eliovson
Design, typesetting and photolithography by
Dieter Zimmermann (Pty) Ltd, Johannesburg
Printed by
Dai Nippon Printing Co (Int'l) Ltd, Hong Kong

Endpapers:
Namaqualand Daisies to the hills and beyond.

A leaven of wildness is necessary for the health
of the human spirit . . .
From the forest and wilderness come the tonics and
barks which brace mankind.

THOREAU

Previous books by the same author:

Flowering Shrubs, Trees and Climbers for Southern Africa (1951)
South African Wild Flowers for the Garden (1955)
The Complete Gardening Book (1960)
Discovering Wild Flowers in Southern Africa (1962)
Proteas for Pleasure (1966)
Bulbs for the Gardener (1967)
Gardening the Japanese Way (1970)

Contents

Illustrations

Foreword

It is probable that more South Africans have visited London or Paris than have been to Namaqualand. Yet a visit to Namaqualand at the height of a good flowering season is a unique experience.

Desert plants show many interesting adaptations which enable them to survive in a harsh environment. Some are able to conserve moisture by a reduction in leaf area or by having hard and leathery leaves. A further development is the storage of moisture in thick fleshy stems or, underground, in bulbs, corms or tubers. An entirely different adaptation is towards the annual habit, which enables a species to survive with the little rain that falls, germinating, flowering and producing seed within a short time and surviving the long dry season in the form of seeds, which may remain dormant for several years.

In no other country has an assemblage of annual desert plants developed to compare with the variety and magnificence found in Namaqualand. When to these are added the great assortment of succulent plants and the scattered desert shrubs, small trees and grasses, the result is an extremely rich, varied and fascinating flora.

Good accounts have been published dealing with certain groups of succulents and they have been further publicised by societies all over the world devoted to such plants. But practically nothing of a popular nature has been written about our desert annuals. A great deal of basic research has been done and probably few undescribed species remain to be discovered, but the information must be sought in Herbaria and scientific texts. A few of the Namaqualand daisies are popular as garden subjects and recently, with improved roads and the occasional pictures in illustrated magazines, interest in the area and its flora is increasing.

The need for more information, attractively presented and authoritative, has now been met by Mrs. Eliovson, who is well known for her informative books on gardening and botanical subjects, including several dealing with our indigenous flora. We are grateful to her for having the pioneering spirit and the talents required to produce a publication of this standard.

A great deal of hard work has gone into the production of this book. Thousands of miles have been travelled over a period of several years to find the flowers at their best because the season is short and the different species vary from year to year. To capture the fleeting beauty of the flowers requires technical skill and perserverance. In addition, voucher specimens must be prepared for later identification in a Herbarium.

The result is an attractively illustrated account, fully representative of the most interesting and colourful species of Namaqualand, covering both annual and perennial plants in their natural surroundings, which will undoubtedly give a great deal of pleasure to a wide spectrum of readers.

L. E. CODD
Director: BOTANICAL RESEARCH INSTITUTE

Acknowledgements

I wish to express my appreciation for the assistance received from the Botanical Research Institute, Pretoria, under the direction of Dr. L. E. Codd, during the preparation of this book, and should like to acknowledge the help of the following botanists in particular:

Dr. O. A. Leistner, Curator of the National Herbarium; Mrs. E. van Hoepen; Mrs. A. A. Mauve; Miss M. Welman; Mrs. Magda du Toit; Mrs. Lilian du Toit; Miss E. Retief; Dr. I. Verdoorn and Mr. E. A. Marshall. The information received from Miss M. D. Gunn, with regard to the early botanists, is warmly appreciated.

My gratitude is due also to the following botanists from the Cape, who have assisted me in naming plants:

Dr. Peter Goldblatt of the University of Cape Town; Miss W. F. Barker and Dr. J. P. Rourke of the Compton Herbarium, Kirstenbosch Botanic Garden; Professor H. B. Rycroft, Director of the National Botanic Gardens; Mr. E. G. H. Oliver and Miss Mary Thompson of the Botanical Research Institute in Stellenbosch and Mr. Ion Williams of Hermanus.

I am grateful to my friends in Springbok, Piet van Heerde, Dave Smith and Mr. du Moulin, who showed me places of interest where I could photograph special flowers in their natural environment; to my friend Doreen Jeffes, who lent me five of her own transparencies, and to the artist, George Boys, who helped me to design the lay-out of the photographs and the jacket.

To my friends who love Namaqualand, too numerous to mention by name, and who have talked to me of their experiences there, I am happily indebted. And of the people who accompanied me on my four trips to this fascinating part of the world, my thoughts will always be fresh and well-remembered.

Introduction

Although this book has been written for the layman who is interested in the spectacle of the wild flowers of Namaqualand, it has been necessary to provide factual information on the specific plants that are found in this region. Many people like to add to their knowledge of wild flowers by identifying them in the wild, often with the idea of obtaining seeds so that they can plant them in their gardens.

Many of the wild flowers of Namaqualand have already found their way into the gardens of the world. Namaqualand Daisies, (*Dimorphotheca sinuata*), also called African Daisies, are obtainable in seed-shops in many parts of the world, as are seeds of *Arctotis fastuosa*, which has been hybridised overseas, and Terracotta Gazanias (*G. krebsiana*). Succulents from Namaqualand are common in many collections.

The task of writing about the wild flowers of Namaqualand, however, was complicated by the fact that this floral region has not yet been fully investigated botanically and there is no check list of the plants drawn up as yet. Many excursions have been made into Namaqualand since early times, especially by succulent collectors, several of whom have written about the plants found there. There has been no book to date attempting to reveal the beauty of all types of Namaqualand's flora and, in order to do so, the task of research has been considerable, especially as it has been undertaken by a horticulturist, who is an amateur taxonomist and not a trained botanist.

To add to the difficulties of going into print, many of the plant groups or genera represented in Namaqualand are presently been revised by specialists who are writing monographs on each genus. I have been privileged in that these botanists have given me information about the particular species found in Namaqualand, as well as their correct, up-to-date names. Some species have not yet been sorted out or named and, even if their names have been decided upon, must not be published until the monograph is published, or the scientific value of the botanist's work could be affected.

Despite all these botanical difficulties, however, I prepared this book in order to assist the traveller who cannot help becoming interested in the marvellous flowers he observes, as well as the enthusiast who is already knowledgable about genera. I have, therefore, had to rely on my own observations and notes made on my several trips to Namaqualand, and have collected numerous specimens which have been identified chiefly by the botanists at the National Herbarium, Pretoria. My appreciation of the kindly help I have received from the Director, Dr. L. E. Codd, and his staff, can never be sufficiently acknowledged.

Most people think that there are no other plants but daisies in Namaqualand. It has been my desire to reveal the bewildering number of plants that do not belong to the Daisy family in this rich flora. This area is a paradise for showy succulents that bloom from mid-winter to spring and, in fact, Dr. R. A. Dyer told me that the late Dr. L. Bolus had remarked that she thought that there were more members of the *Mesembryanthemum* family in Namaqualand than of the *Compositae*. This debatable subject cannot be concluded before the area is fully surveyed by the Botanical Research Institute, but no-one can deny that this is certainly a land for the lover of succulents.

In selecting the plants to be described in this book, the most beautiful or curious have been given first preference, as these are most likely to be noticed by the visitor. Insignificant or weedy plants have been omitted as far as possible, except in the case of shrubs and trees, which are so scarce in this region. It will probably be a great surprise to note how many bulbous plants are to be found in this area and how many genera are represented here. To include more species than I have done, would be to confuse the reader rather than help him. A list of good reference books for further detailed study will be found at the end of the book. The total flora of Namaqualand can only be listed scientifically when the work of many botanists is correlated in the future. Even at the present time, the number of indigenous species in Namaqualand is double that of the flora of Belgium, and it might well be found to exceed that of far larger floras.

An interesting fact is that new species of plants are being discovered even today in this comparatively little explored territory, so that new knowledge will be added for many years to come. It has been most exciting as an amateur to have collected, quite by accident, a new species of *Charieis* and to have experienced the thrill that must have been felt by early botanists when they first explored this floral paradise. This thrill can still be experienced by anyone who discovers the flowers of Namaqualand for himself and most people fall under their spell during a first visit in the spring.

SIMA ELIOVSON
Johannesburg

PLATE 1

1 The striking candelabra of Hottentotskool, meaning Hottentot's Cabbage (*Trachyandra falcata*) rises above the daisies. It is common in central Namaqualand and its leaves are reputed to have been eaten by the Hottentots.

2 A magnificent clump of Namaqualand Daisies (*Dimorphotheca sinuata*), often called Single Namaqualand Daisies so as to distinguish them from the so-called Doubles.

1

2

1   2

← PLATE 2

1 Many kinds of golden daisies line the national road near Springbok. The foreground shows *Osteospermum*, *Dimorphotheca* and yellow *Senecio*.

2 Orange Ursinias and a dozen other little flowers fringe the wheat fields, wherever the farmer has not ploughed.

PLATE 3 →

1 A fallow field at Voëlklip, ablaze with several kinds of daisies that form a cover for numerous smaller flowers beneath them.

2 The approach to the small town of Springbok is heralded by wild flowers, which overflow on to the pavements.

Diamonds and Daisies

Fascinating Namaqualand is an exciting place for the visitor with special interests, as well as for the layman. It is a plant-lover's paradise and a treasure-house for the mineralogist.

"Rock-hounds" will be enthralled to wander on the site of the old copper mine behind Springbok and pick up beautiful green, copper-bearing stones, even though the workings have now been moved elsewhere. The copper mines at Okiep and Nababeep opened up Namaqualand commercially and the old town of Springbok, established in 1852, is growing rapidly, but still retains its simple, country atmosphere.

Some of the richest alluvial diamond mines in the world occur at Alexander Bay, just south of the mouth of the Orange River, while several others are in production at the mouth of the Buffels River along the sandy coastal belt of Namaqualand. Semi-precious stones and unusual minerals used in industry are also mined extensively, such as Beryl, which is used in steel production, and Spodumene, which is used in television tubes. Many others include Mica, Corundum, Fluospar, Tourmaline and Garnets.

The innumerable visitors who throng the roads leading to Springbok in springtime, however, come to marvel at the daisies that paint the countryside with splendour after brief rains have fallen.

It has been said that the very rocks grow flowers in Namaqualand and this is almost the case on the hillsides around Springbok, where every crevice seems to sprout daisies, curtaining tawny rock faces with gold. The sandy pavements overflow with flowers and the surrounding hills and fields are gilded with orange daisies, so that it is no wonder that everyone is entranced by their unforgettable glory. It is probably only on the golf course at nearby Nababeep that the daisies are not popular, for it takes a long time to retrieve the golf balls that fall into their midst!

Although the mecca for flower-pilgrims is Springbok, many people will say that they have been to Namaqualand to see the flowers when they have travelled no further north from Cape Town than Van Rhynsdorp. As wonderful as are the spring flowers of the south-western Cape, those of Namaqualand have a brilliance and profligate bounty that is one of the wonders of the world of nature. They bloom a few weeks earlier than those of the south-western Cape, as the climate is drier and warmer, so that one should try to see them at the beginning of the season and then retrace one's steps southwards in order to enjoy each stage in the unfolding of springtime's floral beauty.

PLATE 4

The very rocks sprout flowers in the hills around Springbok. The colour is provided here by the tall Early Morning Daisy (*Osteospermum hyoseroides*) that opens early, but also closes early in the afternoon, rolling back its petals in so doing.

Flower-lovers often have difficulty in trying to decide where the flowers of Namaqualand begin and where those of the south-western Cape come to an end, especially as some overlap into each area, while others extend into the Kalahari and the semi-desert Karoo.

Many of the Namaqualand flowers extend into the drier districts of the S.W. Cape such as Van Rhynsdorp and Clanwilliam, or even into the Peninsula and the eastern Cape Province, but they must have been collected or seen growing wild in Namaqualand in order to be known as part of the flora of Namaqualand.

The divisional boundary of Namaqualand is the floral area which has been covered in this book and no plant which does not occur wild within these borders has been included. The man-made boundaries defined on the Magisterial District map of the 1st April, 1969, which has been reproduced here, are not as arbitary as they might seem, for the flora changes visibly as one enters this area, while plant geography, geology and the topography of the countryside combine to support the recognition of a special Namaqualand flora.

Namaqualand is the land named after the Nama people, who once lived in this north-western corner of the Cape Province, south of the Orange River. It was formerly called Little Namaqualand, as distinct from the larger area north of the Orange River, which was known as Great Namaqualand. These areas were never strictly defined in the past and it should be noted that this southern portion of S.W. Africa, where the greatest concentration of the Nama people live, has been given to them for their exclusive use and renamed Namaland.

Bounded on the west by the Atlantic Ocean and on the north by the Orange River, which half encircles the distinctive arid region known as the Richtersveld, Namaqualand melts eastwards into the romantically named area marked vaguely on old maps as Bushmanland. The boundary is drawn a few kilometres west of the little town of Pofadder. The southern divisional boundary is marked clearly on the national road, about 32 kilometres (20 miles) south of the village of Garies.

The Cape Region is classified as one of the world's six Floral Kingdoms because it has a unique flora that is one of the richest on earth. The so-called Cape Flora covers the coastal zone from Clanwilliam in the west to Port Elizabeth in the east. Namaqualand's flora, however, belongs to the category dealing with the tropical flora of the ancient world, known as the Palaeotropical Kingdom.

When one leaves the south-western Cape and drives along the main road towards Namaqualand, the terrain itself marks the transition between these two floras, despite a certain amount of overlapping of the flowers that bloom at the roadsides. Climate is recognised as one of the most important forces in controlling plant distribution and this changes considerably, becoming ever drier and warmer as one travels northwards.

Once one leaves Van Rhynsdorp and the huge flat-topped Bokkeveld Mountain to the east of the main road, one seems to be on the edge of Namaqualand, for the landscape changes visibly from rich, lush farmland, becoming increasingly drier and emptier, so that there is almost a no-man's land between the small towns of Nuwerus and Bitterfontein. The fact that the railroad ends at Bitterfontein seems appropriate from both a geographical and floral point of view.

There are patches of colour studding the large open plains from Nuwerus to Garies, but these are formed by the glittering flowers of succulent Vygies or Mesembs, rather than by the daisies that characterise Namaqualand's flora. One sees outcrops of the familiar flowers of Namaqualand here and there, such as yellow Button Flowers (*Cotula*), blue *Felicia*, orange *Osteospermum* or chartreuse Duikerblomme (*Grielum*), but it is mainly the shining cushions of orange, purple, red, white and yellow of the Mesembs that catch the eye when one is travelling along the road.

The true spectacle of Namaqualand's springtime flora generally begins around Kamieskroon. Here it is that one may see fields of orange and yellow, punctuated by blues and purples, that dazzle the eye and form one of the natural floral wonders of the world. Seldom in the world can one see a mountain which is splashed with orange to the highest tip, or a valley which is a solid carpet of gold, coppery-orange or daffodil-yellow daisies. And when one steps into a field to look more closely, there are layers of smaller plants growing amongst them and beneath them. A field of orange *Ursinia* or *Osteospermum* may be undercarpeted with short yellow *Oxalis;* another may be a mixture of orange Single Namaqualand Daisies (*Dimorphotheca sinuata*) and yellow *Senecio*, with a sprinkling of dwarf azure Felicias or sky-blue Sporries (*Heliophila*). Between them may be many fascinating little bulbous plants like cerise Painted Petals (*Lapeirousia*) or dwarf carmine *Babiana*, as well as magenta *Pelargoniums*. I have counted no less than 22 species of plants in a single field south of Springbok. These are:

ANNUALS	*Arctotis fastuosa* (Double Namaqualand Daisy)
	Charieis sp.nov.
	Cotula barbata (Button Flowers)
	Dimorphotheca sinuata (Namaqualand Daisies)
	Felicia tenella
	Grielum humifusum (Platdoring)
	Gorteria diffusa subsp.*diffusa* (Beetle Daisy)
	Heliophila coronopifolia (Blue Flax, Sporries)
	Osteospermum hyoseroides (Early Morning Daisy)
	Senecio inaequidens
PERENNIALS	*Arctotheca calendula* (Cape Weed)
	Gazania krebsiana (Terracotta Gazania)
	Osteospermum sinuatum
	Pelargonium incrassatum (Namaqualand Beauty)
BULBOUS PLANTS	*Babiana geniculata*
	Bulbine alooides
	Homeria miniata
	Lapeirousia silenoides (Springbok Painted Petals)
	Oxalis copiosa
	Trachyandra falcata (Hottentotskool)
SUCCULENTS	*Conicosia pugioniformis* (Varkwortel)
	Tetragonia fruticosa (Kinkelbos)

As one travels northwards, there may be glittering masses of low amethyst Mesembs, while, in some places, dark swathes of bushy mulberry-purple Mesembs extend long fingers of rich colour into greyish-blue scrub. Huge silvery-yellow *Conicosias* wink towards the sun, atop mounds of fleshy leaves, attracting myriads of black midges or buzzing flies that are covered in pollen. They are also attracted to the unwary visitor who tries to photograph the silken, shimmering blooms, forcing him to leave the vicinity very quickly.

The topography and geology of Namaqualand are varied and the rainfall varies accordingly. The long sandy coastal belt that stretches from Alexander Bay to the fishing village of Hondeklip Bay, is known as the Sandveld. This extends inland for some 32 kilometres (20 miles) and its formation is due to the action of the cold Benguella current and of wind. It is extremely arid, with a meagre rainfall of 20-100 mm annually. The red colour of the sand is produced by a coating of iron oxides on the sand grains, and in places where this is washed off or worn off by plant growth, the sand becomes grey or white. The vegetation of the Sandveld is composed largely of succulents, including Euphorbias, Aloes and Mesembs. Sea mists and fogs are sometimes sufficient to supply moisture to succulents where rain never seems to fall. Nevertheless, even in this inhospitable terrain fleeting annuals appear in occasional patches. One may find a vast stretch of blue Sporries (*Heliophila*) casting a cerulean haze over the burning red sand or sporadic glittering sheets of pale yellow *Grielum*.

Namaqualand is situated in what geologists term the Central portion of the Nama system, which extends from Garies in the south to Rehoboth in S.W. Africa. The granite rocks of Namaqualand are among the oldest formations in southern Africa and classed as "Old Granite", which is noted for its abundance and variety of mineral deposits, while that of the S.W. Cape is massed with the folded strata of "Young Granite". The rugged central portion of Namaqualand has an escarpment running from north to south, in contrast to the east-west escarpments of the Cape coastal mountains.

Rising gradually to the central area, known as the Hardeveld, one finds rocky hills, cradling enormous wheat-fields between them, which lap like waves up to the foothills. When these lie fallow, they become a blaze of wild flowers, which often border the green sprouting wheat with bands of orange. The display depends on the rainfall, which varies from 50 mm annually to three times that amount.

The long range of mountains to the west of Springbok, known as the Spektakelberg, has a rich and varied flora as it receives more rainfall. The impressive Spektakel Pass marks the sharp descent from the higher altitude of Springbok to the lower plains of the Sandveld.

The numerous conical hills around Springbok are flower-strewn and rich in many types of plants. Wherever there is a kloof or cleft in the mountains, where the smallest trickle of water flows, one may find moist ledges with choice plants like yellow *Romulea* or purple *Diascia*, or be enchanted by dainty white *Moraeas* and delicate pink *Oxalis* among the ubiquitous daisies.

The high mountainous region of Namaqualand starts in the Kamiesberg, east of Kamieskroon, where many lovely flowers may be seen. This comparatively well-watered range has an evergreen, bushy flora that resembles that of the macchia or

4

ORANGE RIVER
MOUTH

Alexander
Bay

29°

State
Alluvial
Diggings
for
Diamonds

RICHTERSVELD

SOUTH
WEST
AFRICA

N

AFRICA

Holgat River

Stinkfontein

VIOOLSDRIF

NAMAQUALAND

Komma River

Port
Nolloth

Oograbies

P.N.
Reserve

STEINKOPF

Brak River

GOODHOUSE

Pella

Buffels
River
Mouth

BUSHMANLAND

Aggeneys

KLEINSEE
Diamond
Mines

Kamaggas

NABABEEP

Concordia

Bloemhoek

Pofadder

Game
Reserve

SPEKTAKEL
PASS

Drie
Rivier

OKIEP

Koperberg
AIRPORT

WILD
FLOWER
RESERVE

30°

SPRINGBOK

Lammerhoek

Klipfontein

Voël-
klip

Droedap

KENHARDT
DIVISION

Koornhuis

Zwart Uintjies River

Kamee.lboom

Onder
Gamoes

Soebatsfontein
Skilpad
Grootvlei

KAMIES
BERG
RANGE

Gamoep

Hondeklip
Bay

KAMIESKROON

Avontuur

Lelie-
fontein

DIVISIONAL BOUNDARY

CALVINIA
DIVISION

ATLANTIC
OCEAN

GARIES

Groen River

Brak River

NAMAQUALAND

National Road to
Cape Town

To Bitterfontein

VAN RHYNSDORP
DIVISION

N to S (longest part) ~ 320 km (200m)
E to W (widest part) ~ 225 km (140m)

31°

"fynbos" of the S.W. Cape, also known as "Cape Scrub". The only species of *Erica* and *Protea* found in Namaqualand occur here. This is the home of many plant treasures, such as the tiny violet or yellow *Lachenalia*, daffodil-coloured *Bulbinella*, rare *Gladiolus equitans* and the tiny white *Androcymbium* that lies scattered on the hard ground like little pieces of paper. Aloes crown the rocky crests and millions of white Sporries (*Heliophila*) lie like fields of snow among the hills, forming a luxuriant feast for the black karakul sheep that graze among them.

The mountains in Namaqualand are no less fascinating than the flowers. It is when driving towards Kamieskroon that one sees the first conical mountain with the twisted stone top-knot that gives the little village its name. Kamieskroon refers to the stony crown on the mountain-peak and is derived from a Nama word "Kameeras", meaning "bundle mountain". From this point onwards, the mountains become ever more typically conical, more pyramidal, more rock-studded and more desolate as one approaches the Orange River. The massive, dome-like granite tops of some of the mountains have a pinkish hue near Springbok and Steinkopf, while those seen beyond the border town of Vioolsdrift form a band of blues, receding in misty planes to the horizon.

There is a level sandy plain to the east of Springbok, that stretches for a long distance to Pofadder and beyond. When a little rain has fallen, one may see vast fields of apricot and cream daisies (*Arctotis canescens*), almost waist-high where conditions are good. Between these bushy plants are small treasures like blue and pink *Manulea*, delicate washed blue cups of *Wahlenbergia* and the rich royal purple of the Karoo Violet (*Aptosimum*). Carpets of brilliant yellow Gazanias hug the ground and, among stony outcrops, one may be fortunate to see the lovely pink Desert Rose (*Hermannia*).

Where the flowers do not bloom, this is a desolate, dry semi-desert, broken only by occasional drought-resistant small bushes, tufts of Bushman Grass and drifts of dotted succulents. There is no shade from indigenous trees, so that sheep crowd together, tucking their heads underneath one another's necks in order to seek relief from the midday heat.

Travelling north from Springbok to Vioolsdrift, the flora and the countryside together become ever more stark. Brilliant annuals like orange Gousblom (*Arctotis fastuosa*) and purple Mesembs continue to line the roadsides, but the tall Kokerboom punctuates the northern slopes of the hills, forming dramatic silhouettes on the skyline. In isolated places near Steinkopf, the rocky knolls are peopled with the strange succulent plants known as the Half-mens (*Pachypodium namaquanum*). Tall, slender, with a tuft of leaves at the top that resembles a nodding head, they stand guard over the barren valleys, like eerie watchers over a moon-landscape. Many are thought to be centuries old, for they are coverd with long spines and every ring is supposed to represent a year of growth.

Travelling north-west under the bend of the Orange River, one enters the dry, rocky highlands known as the Richtersveld, where few people go. This is the home of numerous succulent plants that are adapted to withstand great drought and fierce heat. Spiky Euphorbias, numerous thorny bushes, dwarf perennials and succulents survive the hostile climate, where some of Namaqualand's curious plants, like the Bushman's Candle, may be seen. Geologists have named this granite region the Gariep System, refering to the old Nama name for the Orange River.

Trees are scarce in Namaqualand and mostly to be found in dry, sandy river-beds, where Acacias are most common. Other trees may be seen, which are all adapted to withstand drought. In general, the entire vegetation of Namaqualand is adapted to withstand drought and is of a semi-desert character, with sparse, low shrublets that have thorns or thin leaves to decrease loss of moisture. Succulents, bulbous and tuberous plants that store moisture are typical, as well as the ephemeral annuals, chiefly of the Daisy family, that bloom easily in dry conditions.

Wild flowers all over the world do not germinate regularly every year and this is a device to ensure that they will propagate their kind, otherwise the whole species could be exterminated during an unfavourable season. Their seed also has the ability to lie dormant for more than one season, awaiting suitable conditions for growth. To visit Namaqualand in springtime is to become an addict who wants to return again, for every year is different and the same flowers seldom bloom twice in exactly the same place.

Namaqualand has much to offer the enthusiast, whether it is the land itself with its mineral treasures or its plants, both bizarre and beautiful.

TABLE OF AVERAGE RAINFALL IN NAMAQUALAND

January	– 4 mm	July	– 26 mm
February	– 7 mm	August	– 24 mm
March	– 10 mm	September	– 15 mm
April	– 14 mm	October	– 9 mm
May	– 23 mm	November	– 6 mm
June	– 26 mm	December	– 4 mm

Total 162 mm annually (a little over 6 ins).

At Springbok, the Average Rainfall per year is 224 mm (about 9 ins).

AVERAGE TEMPERATURE TABLE

Taken at Okiep. (Altitude 927 metres) Springbok is about the same altitude.
About 2 days of frost occur in winter.

	Maximum	Minimum
January	30,5°C	15,4°C
February	30,5°C	15,9°C
March	28,9°C	14,7°C
April	25,1°C	12,0°C
May	21,0°C	9,2°C
June	18,3°C	7,4°C
July	16,6°C	5,6°C
August	19,1°C	6,7°C
September	20,8°C	7,5°C
October	24,7°C	9,9°C
November	26,9°C	12,1°C
December	29,3°C	14,1°C
Annual Average 24,3°C	10,9°C	68°F = 20°C.

7

Planning a Trip to Namaqualand

An arid land for most of the year, Namaqualand becomes a wonderland for a few weeks in early spring, when the veld bursts into a miracle of colour that is unrivalled in any part of the world for its spectacle.

It is extraordinary how one can go again and again to this remote semi-desert region and be fascinated by it in a different way each time. When I think back on my several springtime visits, I remember glorious experiences that include exciting new discoveries, stunning and extravagant splashes of colour spreading over the valleys and roadsides of an otherwise gaunt countryside, together with the marvel of the incredible ability of small and delicate flowers to survive and bloom in hostile surroundings, uncared for by anyone except nature itself. Because wild flowers do not all set seed regularly every year, it becomes exciting to hunt for flower spectacles in different places on different visits, for they are never quite the same every year. This makes successive trips to Namaqualand more interesting and always refreshingly new.

I have known and loved Namaqualand for close on twenty years, ever since I first visited it in September, 1952, when Springbok, the small town that is also its capital, was one hundred years old. It was an exciting adventure then, and the trip remains so today, although the roads have improved and the journey has become easy. The main road is now tarred all the way from Cape Town and bus-loads of tourists leave the city each weekend during the flower season. Daily flights from Cape Town to Springbok were inaugurated during the spring of 1967, so that is now possible to fly there regularly during the week and hire a car for touring in the area.

Despite its flower fame, Namaqualand is not visited extensively by holidaymakers, but mainly by those interested in flowers, minerals or photography. It is not necessary to be knowledgeable about wild flowers in order to enjoy a visit to Namaqualand, however, for anyone who is aware of beautiful surroundings must be impressed by what this unusual part of the country has to offer.

Businessmen, who visit the area to follow their interests in the diamond or copper mines or in trading, are made conscious of the flowers in springtime, for everyone in Namaqualand "talks flowers" from mid-August to mid-September. Every shop-owner or garage mechanic will have earnest conversations about where to go to see the best tracts of flowers, for they vary from spring to spring and this always gives one fresh interest in successive years. The best time to see the flowers also varies, for they may

PLATE 5

1 In Bushmanland, when the flowers do not come, there is a nostalgic beauty in the receding mountains and the desolate plains, which await the touch of rain to bring them to life.

2 The most brilliant of all the Arctotis, *A. fastuosa*, the so-called Double or Bitter Namaqualand Daisy, that was formerly known as *Venidium*, covers great tracts of land, growing waist-high whenever the rains are good.

1

2

1

2

3

4

PLATE 6

1 *Felicia tenella* is a dwarf annual that washes the veld pale blue with millions of little daisies.

2 A yellow-centred *Charieis,* as yet un-named, flaunts its rich blue flowers in the fields around Springbok.

3/4 *Felicia namaquana* has large flowers that droop backwards with the weight of the florets. Blue and mauve colour forms adorn the hot red sands near Aggeneys.

PLATE 7 →

Myriads of Button Flowers, *Cotula barbata,* like daisies with their petals pulled off, gild an entire landscape, seeding themselves in thatched roofs, to bloom wherever they can gain a tiny foothold.

start a week earlier or remain in bloom a few weeks later, depending on the rainfall, while hot winds in early September could shrivel the delicate flowers even when they are at their peak.

The precursor of a good season is a brief fall of rain during late February, March or April, which is enough to start life in the seeds and bulbs that lie in their millions awaiting favourable circumstances for growth. Normal average rainfall in the area around Springbok and Okiep is about 160 mm a year (about 6"), but may reach three times that amount in a good year. Exceptional years occur about once in a decade, but there is seldom a spring without exceptional sights, no matter how scattered. It is most unusual that there will be such a bad year that the journey will not be worthwhile, for there is always something to be enjoyed even if the local people say that it is not a good year.

A second rainfall during the winter months of June or July will ensure a reasonably good display of flowers in the spring. It is generally necessary to check that all is well as August approaches, however, if one is to be quite sure that the trip will be rewarding. Unless one is informed positively by reliable sources that severe drought has spoilt the flowers, however, one should endeavour to undertake the journey, for, even in an ordinary year, the sights are so amazing that only a most jaded person could be discontented. Even in a drought-stricken year as in 1971, when good rains fell as late as July, and the total rainfall from March to July was 56 mm, there was a fabulous display of flowers in the area around Springbok, but the peak period of bloom was about a fortnight or three weeks later than usual.

One may marvel at the scant rainfall needed to turn this drought-stricken region into a flower garden, but it must be remembered that sea-mists roll in from the nearby Atlantic Ocean, bringing moisture to the plants, while nightly dews settling on them are sufficient to keep many succulents alive throughout the year.

A Karakul sheep farmer, situated about 30 kilometres east of Springbok, said that he had received 100 mm of rain over four years, and that the most he had received in the previous few seasons was about 20 mm. He had a huge farm, for each of his sheep needed 24 ha (about 60 acres) of grazing, yet when we drove with him in his truck or "bakkie" across the veld, we were rewarded with the unbelievable sight of golden mounds of daisies studding the red earth to the horizon. Between the coarsely rugged orange daisies (*Arctotis canescens*) and the silky yellow flowers of *Conicosia*, the coal-black Karakul sheep moved slowly in a line, silently and obliviously chewing at the golden feast around them.

In areas with more rainfall there is a tall, robust orange daisy (*Arctotis fastuosa*) that grows waist-high during a good season. It is possible to farm with cattle in such places, but when the cows graze on these plants the milk becomes tainted, so that the flowers are referred to as the Bitter Gousblom or the Bitter Double Namaqualand Daisy.

In rocky kloofs in the mountains there may be tiny running streams where one may

PLATE 8

Glorious Gazanias open to the midday sun. The large Terracotta Gazania, *G. krebsiana*, with its brilliant colouring and intriguing markings, is one of the loveliest, but the dwarf yellow *G. lichtensteinii*, with tiny black spots, adds a rich splendour to the barren earth in dry places.

9

find different types of flowers from those that blaze in the open fields. Many are small treasures that are of special interest to plant-lovers and include numerous bulbs and perennials. The amount of water available, therefore, affects the kind of plants that one may find in different places. Even in the depressions on road verges, flowers gather more thickly because of the extra moisture that collects there, apart from the fact that the soil has been cleared and turned during road-cleaning operations.

The traveller who wants to see wild flowers in Namaqualand at their best must try to assess the amount of rainfall that has fallen in autumn and winter. This information can be obtained from the Weather Bureau, from Publicity Associations or from private residents. He can then judge exactly when to visit the area, besides deciding on how long to spend there and exactly where to go.

Bearing in mind all the vagaries of the rainfall and its effect on the flowers, as noted above, it seems that the safest period to visit the flower fields is from the middle of August until the first week in September. In a cold season it would be inadvisable to go before mid-August, although the flowers have been known to start blooming at the beginning of August, particularly in the warmer, northern parts near the Orange River. To go after about the 6th of September would normally be to court disappointment, as the main spectacle would then be nearly over and the season could be cut short abruptly after a day or two of hot, dry winds. Very late rains, however, would make the season last until the end of September.

One should aim to spend at least three days in Namaqualand, even if one is not a flower enthusiast, while the flower-lover could spend anything from a week or more in the area without risk of boredom or repetition. One must remember that there may be cloudy days in early spring, even accompanied by light, misty rain, and that the flowers do not open except in bright sunshine during the warmest part of the day. One can seldom expect to see a really colourful stretch of flowers before 10 or 10.30 in the morning or after 4 in the afternoon, so that the hours of flower-hunting are curtailed. It is advisable to be out during the midday hours and to take refreshments in the car, so as not to waste time having a formal lunch. It is true that one can use the hours when there is little flower colour for driving long distances, but one should not set out too early as part of the pleasure in driving is to see the flowers, while they always appear to be different on the return journey. The flowers are best seen when travelling with the sun behind one, for many of the daisies and buttercup-like blooms are then fully open towards the traveller as they turn towards the sun's warmth. When one approaches from behind the flowers, their brilliant effect is inevitably dulled.

Then, also, it is necessary to know the habits of some plants. Daisies like those of *Ursinia* will open throughout the warm hours of the day, while others, like *Osteospermum*, will open even on dull mornings, but will roll up firmly after midday and remain fully closed for the afternoon. On the other hand, some daisies, like *Felicia*, and succulents, like *Conicosia*, open fully only at midday and remain open until later in the afternoon.

Exact advice on where to go will depend on many things, based on how long one stays in the area as well as on one's headquarters. It is most important to spend at least two nights at Springbok, for one can make half-day or whole-day trips to the west at Port Nolloth, to the east towards to Pofadder or as far north as Vioolsdrift at the Orange River, with its gaunt moon-landscape terrain.

A day can be spent driving south for about 40 kilometres from Springbok to Kamies-kroon, with the sole purpose of visiting the farms of Skilpad and Grootvlei, which are a few miles to the west of the main road. This is one of the best areas in which to see flower spectacles, for the farmers have left several wheat fields fallow so that the daisies and other plants have taken over, spreading a riot of colour to the base of the surround-ing conical hills. Horses and donkeys stroll nonchalantly in this paradise, looking up only to stare at the photographer or bedevil the tourist who might have forgotten to close the farm gates between the fields and must spend his time and energy chasing the strays back to their rightful domain. In a drought-stricken year, however, the flowers may not appear with their usual exuberance.

Turning to the east through the village of Kamieskroon, there is a road into the moun-tains which is sprinkled with interesting plants, like Lachenalias and Bulbinellas, as well as beautiful fields of flowers, especially White Sporries (*Heliophila*) that lie like drifts of snow on the hillsides. One could spend at least a day exploring the Kamiesberg mountains, should there be time, and stop overnight at Kamieskroon.

The best way to reach Namaqualand is by road from Cape Town and then to return along the same route, a matter of about 6 or 7 hours driving each way. At best it is advisable to break the journey to Springbok, staying overnight at Clanwilliam, Van Rhynsdorp or Klawer. Although it is possible to drive the full distance (560 kilometres or 350 miles) in a day, this will allow little time to enjoy the scenery and flowers that may entice one to stop and look more closely at them. If one is approaching Nama-qualand from other provinces, without going as far as Cape Town, then the best route from the national road is from Victoria West through Calvinia to Van Rhynsdorp, linking up at that point with the road from Cape Town. The road from Upington to Springbok through the northern Cape is not good.

While much can be seen along the main roads through Namaqualand, sorties along the sandy country roads to places like Hondeklip Bay, Komaggas or Concordia, as well as along the many tracks that lead to isolated farm houses, can result in discoveries that will thrill the plant-lover and would inspire the artist.

Namaqualand is a place where it is best to have one's own car and be free to take one's time. It is ideal for the caravan-owner or the camper who is not forced to return to his hotel by nightfall. Yet the visitor with sophisticated tastes can also ramble through the countryside, provided that he reserves his accommodation well in advance and is prepared to drive to and fro in order to return to his base each evening.

Hotels in Namaqualand are limited in capacity, for they deal only with day to day requirements throughout the year. A few are geared to cope with the sudden influx of flower hunters during spring, but last-minute accommodation is often a problem, especially on weekends.

Today's traveller expects all modern conveniences in hotels, even in out-of-the-way places, and progress has certainly brought these to Namaqualand. To look back on the conditions in the less popular hotels or boarding houses only two decades ago, brings back amusing memories of thin, un-lined curtains scarcely covering the sashed windows; cold water in the ewer standing in a basin on a marble-topped dressing table; fires having to be lit under the bath in order to obtain hot water for the small amount of water allowed for bathing, as well as quite unspeakable toilet arrangements. These memories fade into the background, however, being remembered only as part of the fun of exploring places off the beaten track and making one feel like one of the early explorers and botanists who expected to put up with hardships. While not altogether regretted, little experiences like these are now only for those who camp in the veld, for the hotels have now improved out of all recognition.

Scarcity of water has always been a problem in Namaqualand, while much of the water in the district is brackish and unfit for drinking. Bath water is often limited during hot, dry periods, especially during the fiercely hot dry summers. Springtime tourists, however, generally strike very cool weather in the early mornings and evenings, with strong sunshine and heat for about five hours in the middle of the day, during which time the daisies unfurl and bask in the sun's warmth. One needs a woollen jersey and a coat or jacket for the early mornings and evenings, for it can be surprisingly cold during the night, but one must be prepared to peel off warm clothing during the middle of the day and wear a thin cotton blouse or shirt underneath in order to be comfortable at all times.

A trip to Namaqualand is an adventure even in this modern age, but it remains the kind of adventure that is feasible and not dangerous in any way. There is the excitement of the unknown and unusual, with places far from civilisation and untouched by progress, with quaint names redolent of the discoveries of early explorers and of the primitive people who were found there more than a century ago.

Even if one is not a plant-lover, a trip to the flower fields of Namaqualand will be a rewarding experience to anyone with an appreciation of beauty in nature.

Suggested Itineraries

These suggestions for trips within Namaqualand are based on the length of time that tourists have available to spend in the area. One should travel as far as Springbok even if only one day is available. Springbok should be made one's headquarters as it is possible to take at least four separate day-trips in its vicinity and return to base each evening.

Nababeep is very close to Springbok and can be used as an alternative base if accommodation in Springbok is unobtainable.

Kamieskroon is another pivot for the southern portion of Namaqualand, from where one can drive eastwards into the Kamiesberg for a day, or west of the national road in order to enjoy the spectacular fields of flowers that are generally to be seen at Grootvlei and Skilpad.

These schedules can be followed strictly only if the days are sunny. The annual daisies and Mesembs will not open on dull days and their brilliance will be lost. On such days, one could seek out the dramatic plants such as the Kokerboom and Half-mens that do not depend on sunshine for their interest.

It is always best to try and stay an extra day or two, in order to allow for dull weather and so make the most of one's visit. The weather is often changeable, being cloudy in the morning and clearing to fine, sunny weather at midday, so that it is impossible to predict ahead too far. On the whole, the spring days are clear and hot, while cloudy weather does not generally last for longer than a couple of days.

The air is particularly clear and pure in Namaqualand, very dry and without a vestige of smog. The nights are bright with every star crisply outlined and the dawn is like a lucent crystal. Mist occurs near the coast, but is quickly dissipated in the dry inland air.

The schedules outlined in the following trips, which are planned to cover 1 to 7 days, are arbitrary and drawn up purely as a personal recommendation. They should be regarded as advice given to friends in the hope that it will enable them to see as much of Namaqualand's attractions as possible in a limited time.

As the spring flowers are in bloom earlier than those of the S.W. Cape, it is suggested that one should visit Namaqualand first and then follow on with a floral tour southwards, ending up in the Cape Peninsula and Caledon, which are at their best from mid to later September. The local Cape Wild Flower Shows usually take place during the second half of September and may be taken as a guide as to when the flowers in each area will be at their best, for they vary each year. Exact dates can be assertained from the Cape Publicity Association.

ONE-DAY TRIP

Daily weekday flights from Cape Town leave at 8.30 a.m. to arrive in Springbok about 10 a.m. A hired car could meet the plane and one could drive on the roads around

Springbok to see the fields of flowers until after 3 p.m., returning to the airport in time to catch the late afternoon plane back to Cape Town.

A day trip is recommended only for the hurried tourist, for the time is too short to allow for more than a cursory appraisal. This trip may be enjoyed to the full only on a sunny day and if the flowers are plentiful around Springbok.

TWO-DAY TRIP

Driving up to Springbok on the first day, one will see the flowers in bloom at the sides of the road from Kamieskroon to Springbok, but one will have to look back to enjoy them, for they face northwards towards the sun and will not be at their best when seen on approaching them from the south.

The first morning should be spent driving to the foot of the Spektakel Pass and then retracing one's steps to Springbok. If one can spare the time, drive to the golf course at Nababeep, which is covered in wild flowers, in the afternoon.

The second day should be spent driving slowly southwards to Kamieskroon, aiming to linger at Grootvlei from midday before continuing southwards in the afternoon in order to leave Namaqualand. It will take about 1 hour to drive between Springbok and Kamieskroon, but one can spend time at private farms south of Springbok from 10 a.m. to 11 a.m. before proceeding southwards.

THREE-DAY TRIP

Spend the first day on Spektakel Pass, continuing to Port Nolloth if there is time. Spend the second day driving to Nababeep and then continue northwards to Steinkopf or Vioolsdrift. The third day should be spent returning to Kamieskroon, as outlined above.

FOUR-DAY TRIP

  1st day: Spektakel Pass and beyond.
  2nd day: Nababeep and Steinkopf or Vioolsdrift.
  3rd day: Drive to Agenneys or road to Pella and back.
  4th day: Kamieskroon and Grootvlei.

FIVE-DAY TRIP

  1st day: Spektakel Pass and beyond.
  2nd day: Nababeep and Steinkopf.
  3rd day: Agenneys or road to Pella and back.
  4th day: Trip to Soebatsfontein and Hondeklip Bay, returning to sleep at Kamieskroon.
  5th day: Grootvlei and Skilpad.

SIX-DAY TRIP

  1st day: Spektakel Pass and beyond.
  2nd day: Nababeep and Steinkopf.

3rd day: Agenneys or road to Pella and back.
4th day: Trip to Soebatsfontein and Hondeklip Bay, returning to sleep at Kamies-
          kroon.
5th day: Grootvlei and Skilpad.
6th day: Drive into Kamiesberg to Leliefontein and Nourivier. Return to Kamies-
          kroon.

SEVEN-DAY TRIP

Stay longer in the Springbok area, visiting private farms within half an hour's drive
south of Springbok, such as the farms of Lammerhoek and Voëlklip. When visiting
private farms, always ask the farmer's permission to drive through the farm, close
gates between the fields and do not turn around in the narrow roads so that the flowers
on the verges are spoilt in any way.

EXTRA DAYS IN THE SPRINGBOK AREA

A drive to Droedap on the road to Gamoep is interesting in a good year, where many
different bulbous plants and annuals may be found. This is a very hot, dry area. One
may plan to go as far as Gamoep and even return over the Kamiesberg to Kamieskroon,
but this is a long drive.

A walk into the hills on the farms south of Springbok, especially at Voelklip, may
reveal a short perennial stream coming down the kloof or cleft in the rocks. One will
be rewarded here with the sight of choice and unusual plants apart from fields of daisies.

A visit to the HESTER MALAN WILD FLOWER RESERVE, near the Springbok airport,
may be arranged. This land, which was presented by the Okiep Copper Mining
Company, covers about 5 200 hectares (6 000 morgen) in extent. It consists of hilly
country, to the east of Springbok, that is covered in wild flowers and where wild
animals, such as Gemsbok (Oryx), Springbok, Duiker, Klipspringer and Baboons
may be seen.

A visit to Van der Stel's first copper shaft in the Koperberg, which has been declared
an Historic Monument, may be made. This is only a few minutes drive east of Springbok, just off the main road to Pofadder. A walk in the hills in this vicinity would enable
one to approach the Kokerboom (*Aloe dichotoma*) that grows near this spot.

The Public Library in Springbok is most interesting, especially for those who would
enjoy seeing the excellent exhibit of gems and minerals of the area. This is situated
alongside the Caravan Park.

Rambling in the Veld

To visit Namaqualand in order to see the flowers, to hunt them with camera, sketch-book or note-book in hand, is to experience the excitement of the collector who seeks treasure trove. There is more to see in the veld in Namaqualand than flowers and plants, however, for one can revel in nature in the yet unspoilt countryside. The inter-dependence of flowers, insects, birds and animals is never more apparent than when sitting still in the veld. At first there is an awareness only of silence, then the sounds of birds and of insects begin to be heard. The bark of a baboon may echo from behind a distant rock, or a timid buck may appear, but one should be up early to see the life that is all around, yet hidden from the eyes of the hurried tourist in a crowded bus or swiftly moving car.

If one has sharp eyes one can walk in the veld only a short distance from the car and see many signs of life between the tufted bushes. The footprints of small buck or the arrowed imprints of stalking birds in the soft sand can be recognised even by the uninitiated. Those who can identify them may even see signs of the little tortoises that are common in these parts and which are so attractive that they have been protected by law. They propel themselves forward with backstroke movements of their thick curving forelimbs, so that they make marks like a miniature tractor along the sloping sandy hillocks before disappearing into the bushes. Tiny lizards with brilliant peacock-blue and green colouring scuttle into rock crevices at the sound of a footfall.

Many of the smaller animals that roam this region come out at night to hunt still lesser creatures, and may be seen only by the traveller who is about at sunrise. The Black-backed or Cape Jackal can be seen in the very early morning as it trots towards rocky places or sheltering bushes where it will rest during the day, or it may be seen in the late afternoon when it begins to prowl. Although it is a scavenger, it also steals the farmers' sheep and preys on small animals, birds or insects. The Silver Jackal or Cape Fox (Silwerjakkals) is a true fox that makes its burrows in this sandy semi-desert, but is seen only rarely. It is also a predator on rodents or insects, but does not kill sheep.

Wild cats roam about at night, hunting small creatures. Of these, the reddish Caracal or African Lynx (Rooikat) may sometimes be seen during the day, but the tawny spotted Serval (Tierboskat) is seldom glimpsed.

Brown Rock Rabbits (Hyrax) or Dassies are common throughout the country and are often the prey of wild cats when they emerge at night to feed on leaves, roots and bulbs. When they bask on warm rocky outcrops during the day, they are often seized by eagles, especially the magnificent Black Eagle. The greyish-brown Cape Hare (Rooipootjie) is also common throughout the country, living on grassland as well as amongst the small Karoo bushes of the dry areas.

The presence of the amusing, long-snouted Antbear (Aardvark) can be felt only when one sees the large holes that it claws in the earth while searching for termites, for it is active only at night and rarely seen during the day.

PLATE 9

A melange of yellow *Senecio abruptus* and orange Namaqualand Daisies (*Dimorpho-theca sinuata*) form incomparable colour harmonies that can only be found in nature.

PLATE 10

Amongst the hills and valleys of the Kamiesberg there are snowy drifts of white Wild Flax or Sporries, where the Karakul sheep munch silently at their luxuriant feast. This is most likely to be a mixture of *Heliophila seselifolia* and *H. namaquana*, but there are several other species that have similar dainty white flowers.

1

2 PLATE 11

1 Blue Sporries or Wild Flax, dancing in the breeze on the red sands of the coastal belt near the Büffels River. This is *Heliophila bulbostyla*, with bright blue flowers, which also occurs elsewhere.

2 *Heliophila coronopifolia* has a distinguishing pale greenish-yellow centre and may be seen near Springbok.

1

2

Small Ground Squirrels frisk about in the sand in the late afternoons and early mornings, fanning out their furry tails as they squat to nibble seeds and bulbs that they hold in their forepaws. They do not seem to resent being watched by motorists who remain in their cars and enjoy their antics. Their companions are sometimes the Suricate or Slender-tailed Meerkat, and the Bushy-tailed Meerkat, also known as the Yellow Mongoose. Both types of animals eat small creatures, insects and birds, but are themselves the prey of eagles and the powerful black and white Ratel or Honey Badger.

Several types of buck live in this region. The name of the capital, Springbok, is a tribute to the lovely Springbuck that drifts over the land and onwards into South West Africa. It is also able to survive in great heat with very little water and exists on the succulent scrub that it nibbles daintly throughout the day. The great grey Oryx, or Gemsbok, with its rapier-like black horns and precise black and white facial markings, also exists in herds in the more distant areas, especially to the north. It may also be seen in the nature reserve near Springbok. The tiny, timid Steenbok, its wide-open ears listening for every sound, darts off into the veld when disturbed in its grazing. Klipspringers abound in the hillsides and leap from rock to rock with sure-footed agility.

Bird life in Namaqualand is prolific. Practised bird-watchers would be especially aware of the variety of birds that wing across the sky or perch in the bushes, but even the layman could not fail to notice some of the distinctive birds that are common in this semi-desert region.

The large Karoo Korhaan struts about restlessly over the veld, searching for insects among the stones, and the huge, spotted, grey Kori Bustard or Gompou stalks rapidly across the open, ready to spring into the air if it is approached too closely. It is matched in size only by the majestic Secretary Bird, ubiquitous on the grassy plains and drier parts of Africa.

Many other birds that are seen commonly in the country also traverse the Namaqualand region. Swifts and swallows fly high overhead, where insects have been swept upwards by hot air currents; the yellow-breasted Bokmakierie trills its onomatopoeic call from Karoo bushes, as well as from perches around farm homesteads, and the grey Karoo Chat and the brown speckled Ant-eating Chat flutter from their perches on ant-heaps and telegraph poles in Namaqualand, as in other parts of the country. Large Black Crows croak noisily in the open fields. In the river beds, especially around the Orange River, pink-footed Egyptian Geese and other waterfowl may be seen roosting or feeding. The Cape Francolin also frequents stream-sides and cackles loudly from under small bushes in the evening and at dawn. The fascinating Social Weaver, that weighs down small trees with its enormous nest cluster, is present in the valley of the Orange River and extends northwards.

Few birds are found exclusively in Namaqualand, being common to all the dry Karoo areas nearby, but some have had the name Namaqua incorporated in their common names, as they are so prevalent in this region. The cinnamon-winged Namaqua

PLATE 12
1 The glowing flowers of *Pelargonium incrassatum* signal attention from afar.
2 *Pelargonium echinatum* flourishes on dry stony hillsides. There are several colour forms.

Dove nests in the Aloes or low bushes of Namaqualand, but also spreads throughout Africa and beyond, especially in dry areas. The long-tailed Namaqua Sandgrouse is an especially loved bird because of its distinctive call that has had its sound recorded in the Afrikaans name of Kelkiewyn. They cluster in their hundreds, and sometimes in their thousands, around waterholes, living mainly in Namaqualand and the Karoo, although they can be seen in all dry western parts of South Africa. The Namaqua Prinia is a rare bird that is confined to the northern Cape and would probably only be spotted by a keen bird-watcher.

The Afrikaans name, Namakwa Suikerbekkie, is given to the Dusky Sunbird, which thrives in the Karoo, Namaqualand and into the dry areas of S.W.A., feeding on the Kokerboom flowers and nesting in prickly Euphorbia plants. Although it has the same elegant curved beak as other Sunbirds, it is drab in comparison to the jewel-like green Malachite Sunbird or the colourful Lesser Double-collared Sunbird. They, too, feed on the nectar of long-tubed flowers like Aloes and Ericas, and it is exciting to see either of these birds through binoculars.

Millions of pollinating insects appear among the open flowers during spring, often worrying the traveller with their buzzing and biting. Midges swarm in the hot air, so that the farmers' labourers tuck large handkerchiefs under their hats to shield their necks and cover their ears. A large fly with iridescent green eyes seeks out the unwary picknicker in the daisy fields, often making him retire to the protection of the glass windows of the car. Glossy black Toktokkie beetles or button-like grey beetles scurry along the ground between the flowers.

Violet colours are said to be attractive to insects and it is true that many of the little bulbs or daisy flowers are coloured a shimmering violet, cerise or pale mauve. Yellow is a colour that has a common attraction for both birds and insects, and the predominant colour of the wild flower fields is yellow or orange. There are very few pure reds in the wild flowers of Namaqualand, the majority being a purplish-red that is thought to be more visible to insects than to birds, although butterflies also visit red and pink flowers. The lovely markings on the lower petals of some bulb flowers, like *Lapeirousia* and *Babiana*, also act as nectar-guides to the insects. *Gorteria* is a daisy with three small petals at the centre that resemble tiny beetles, which is thought to be a device to attract pollinating beetles.

The myriad insects themselves provide food for the birds that visit the Namaqualand area, but the birds also pollinate some flowers, dipping their beaks into tubular flowers like the stiff, reddish-mauve flowers of *Erica plukeneti* or the numerous Aloes that bloom so brilliantly in the driest areas.

And so the role of nature is enacted in this semi-desert area, which supports a wealth of animal and bird life in addition to its spectacular flowers, with each kind enfolding its purpose in the life cycle of the other. It is for the observant naturalist to study and become aware of this interaction of nature while enjoying the floral spectacle.

Tips for Photographers

Flower photography in Namaqualand differs from that in other parts of the country because of the brilliance of the flowers and the bright clear light of this region. The air is very dry and there is no humidity to soften the atmosphere. It is also extremely clear, without the pollution of smoke to cloud the environment.

The flowers themselves are very bright and reflect light as does the light sand on a beach. Yellow flowers are brightest of all, with orange flowers seldom darker. Blues and mauves reflect less light and this is distinctly noticeable when a light-meter is held near them. It is advisable to use a light-meter in order to take readings, as the light is almost twice as bright as the normal light in other sunny parts of the country.

Many of the flowers in Namaqualand open and close at different times. They are usually all at their brightest and best from 11.30 a.m. until 3.30 p.m. After that time the intensity of the light weakens and the flowers lose their glitter. Some begin to close and it is best to desist from taking photographs after 4 p.m.

Some flowers, like *Osteospermum*, open fully as early as 9 a.m. but begin to roll back their petals at 2 p.m. and cannot be photographed later. The *Felicias* seem to unfurl properly only at noon, remaining open until about 4 p.m. *Conicosias* glitter most spectacularly in the afternoon, remaining quite fully open even after 4 o'clock. It is best, therefore, to be ready to take pictures of all spring flowers from about 11.30 in the morning until 3.30 in the afternoon. Some flowers, like *Lachenalias*, do not open or close, while succulents like Aloes are always open. One must make use of this knowledge if time is limited.

Mood pictures of scenery are best when the long shadows cast by the afternoon sun give a third dimension to rocks and mountains and silhouette the fantastic shapes of the Kokerboom. Early morning shots will have the same effect, but it is to be remembered that very early morning colour pictures may have a yellow cast, while late afternoon colour pictures will pick up a warm reddish tone. Unless these effects are desired, it is best to limit one's colour photography to the central hours of the day.

Photographing flowers in Namaqualand will reward one with the greatest pleasure, for it is only when one seeks out the flowers in order to photograph them that one becomes involved with them and is made aware of the fabulous variety of the flora in this part of the country.

Tips on Growing Namaqualand's Flowers

No gardener can visit Namaqualand without wishing to grow some of the flowers he sees in order to recreate a little of its beauty. Many of Namaqualand's annuals have already become popular garden flowers, such as Namaqualand Daisies (*Dimorphotheca sinuata*), Bitter Namaqualand Daisies (*Arctotis fastuosa*), Button Flowers (*Cotula*) and various kinds of *Felicia*. Numerous succulents from Namaqualand have formed an important group for collectors.

There are strict rules protecting wild flowers in order to prevent people from removing them from the wild, but many of their seeds are available in seedshops as well as from the seed-lists of the Kirstenbosch Botanic Garden and the Caledon Gardens in the Cape. Plants are available from nurseries which specialise in indigenous plants, as well as from ordinary nurseries, and are well worth seeking.

Despite the fact that Namaqualand is an arid region, the flowers grow well in ordinary gardens. It should be remembered that the chief planting time is in the autumn, as this is the time of the onset of rain in Namaqualand, as most of its plants bloom in springtime.

The chief requirement of the soil is that drainage should be good and that it should be light, but not over-rich. Compost may be added with safety to both sandy or clay soils. The fact that the annuals will grow in poor soil does not prevent them from thriving in good garden soil and they have been seen in bloom in many countries around the world from England to America and Japan.

Some plants, especially bulbs, require moist soil as they grow in damp places in nature, such as *Romuleas*, *Homerias* and certain types of *Oxalis*, as well as annuals like *Nemesia*, perennials like *Diascia* and *Monopsis*, and succulents like *Crassula natans*. Nevertheless, they will withstand drought and survive in drier conditions, even if they do not flower to their best advantage in adverse circumstances. In fact, it is probably easier to grow Namaqualand's flowers than most other plants and the only danger is that of overwatering, in combination with bad drainage.

Bulbs are particularly prone to rot if they are watered when they become dormant, especially soft bulbs like those of *Lachenalia*. It is best to grow these in pots and remove them quickly to a covered shed as soon as the foliage has ripened, so that there is no danger of the soil becoming wet while they are not growing. They can be kept in the pots in bone-dry soil until the following growing season commences, or they can be taken out of the soil and stored in a dry place until they need to be planted out again in early autumn.

Late March is the time for planting out Namaqualand bulbs and seeds of annuals in South Africa and in the whole southern hemisphere. If they are grown in the northern hemisphere, they can be planted in August in places with mild winters. Although they can stand low temperatures and some frost, they cannot take freezing temperatures and need to be planted in spring in countries with very cold winters.

20

It is best not to broadcast seed in the ground before the weather has turned a little cooler or the plants may produce luxuriant growth that will be affected by early winter night frosts. Namaqualand Daisies, therefore, are better planted during the first week in April than at any other time.

Pots containing small bulbs should be placed in sunny sheltered positions or the foliage may be burned by frost and the subsequent flowering spoilt. As they frequently grow in places shaded by rocks, it does not spoil them to be given some shade. The dappled shade from tree branches will not only prevent pots of bulbs from drying out, but will protect them from the fall of frost, especially in dry, inland areas. It is quite a good idea, therefore, to keep a collection of pots containing Namaqualand bulbs under a spreading tree.

Shrubs require extremely good drainage and a sunny position. They are very drought-resistant and can be placed at the top of a rock garden where they will not be over-watered as drainage will be good automatically.

Succulents require special treatment, depending on their size. On the whole, sandy, well-drained soil suits them best. Water will not hurt them provided that the soil drains rapidly and the atmosphere is dry. Small succulents should be grown in well-drained pots of sandy soil and kept under cover in order to protect them from rain. Large Aloes and Euphorbias can be grown out of doors with safety, provided that the situation is sunny and warm, and must be grown under cover in countries with cold winters.

How to find the name of a Wild Flower

The serious plant collector who requires authentic identification should press specimens and send them to an Herbarium attached to a University or a Botanical Garden. It is not generally known that the public can take advantage of a service provided by the Botanical Research Institute, by having both exotic and indigenous plants named for them, provided that they submit a suitable botanical specimen. One should send them to the National Herbarium, Vermeulen Street, Pretoria, which is under the auspices of the Botanical Research Institute of the Agricultural and Technical Services Section of the Government, or to the branch in Stellenbosch.

Cape and Namaqualand plants in particular can be identified by the Bolus Herbarium, University of Cape Town, and the Compton Herbarium at Kirstenbosch.

No specimens can be picked for collection unless one first obtains a permit from the Department of Nature Conservation, otherwise the amateur can be fined heavily. The rules of the Wild Flower Protection Act governing the Cape should be studied when collecting plants in Namaqualand, but exact instructions will be given on the permit.

HOW TO PRESS SPECIMENS, AMATEUR-STYLE

If one intends to collect specimens of plants for naming, one should carry a small notebook in which to jot down particulars or make rough labels, for one soon forgets details after picking a flower or branch and then moving on.

It is best to number each specimen and then write down the numbers in the notebook and write down the details there. Details to be noted are the date; the distance from the nearest town; the type of locality in which the plants grow, such as sandy, stony, shady or moist; colour and texture of flowers; size of leaves or any other feature that may be lost when the plants are dried.

If one does not own a botanist's press, one can use a light simple makeshift that will be effective and easy to arrange. Try to obtain 2 pieces of cardboard, each large enough to cover a sheet of newspaper that is folded in half. Thick folded newspapers may be used instead of cardboard, but are not quite so easy to handle. Each specimen will have to be inserted into a sheet of folded newspaper, which must then be placed between the cardboard sheets and held in position by two bands made from lengths of elastic. It is easy to slip off the bands while travelling, so as to insert additional specimens, each to be first placed between a separate piece of folded newspaper and stacked in a growing pile.

The flowers must not be allowed to dry out or they shrivel and fade, but should be pressed immediately and spread out so that they may be seen clearly. The newspaper absorbs moisture and assists in drying out quickly. It is essential to put a heavy weight on the specimens in order to press them flat and, until one can return to base and place them under a heavy book or suitcase, one may sit on them in the car!

In order to preserve the specimens properly it is necessary to put the newspapers into fresh sheets each evening as the used newspapers become damp. If this happens, the specimens will become mouldy and useless for identification. One should change the newspapers for at least 7 days, unless one can take the plants to their destination and hand them over to a botanist before then.

The People of Namaqualand and the Plants

The people of Namaqualand are no less interesting than the plants and, even today, there are some who can enthrall one with stories of the early days and of the discoveries and fortunes that were made there. Very little has been written about the pioneers who inhabited this land of adventure and romance. The following notes on the early people in Namaqualand may be of historical interest, especially with regard to their relationship with the plants that they found and used there.

THE NAMA PEOPLE

Namaqualand is an old name for the country in which the Namaqua tribe lived, Namaqua being the plural of Nama. The Little Namaqua lived south of the Orange River in what was formerly called Little Namaqualand and the Great Namaqua lived north of the Orange River in the southern part of South West Africa.

The name Namaqualand is used nowadays to denote only that portion south of the Orange River that was formerly called Little Namaqualand and it should not be confused with Namaland, which is north of the Orange River in South West Africa, where the greatest concentration of the Nama people live today. Some live in the Richtersveld and some as far south as Springbok, but these are no longer purely Nama and tend to be a disappearing race, like the Bushmen.

The Nama are a Hottentot people. They form the best-known of the four divisions of the Hottentot tribe and their language is the best-known of the four Hottentot languages. The Nama language is used today in their schools and publications and has left its mark on the strange-sounding place-names of Namaqualand.

The Hottentots called themselves the "Khoi-khoin", meaning "men par excellence" or "men of men" and applied the name "san" to the Bushmen, who had no collective name of their own. The term "Khoisan" was coined by Schultze to denote the stock to which both Bushmen and Hottentots belong and an interesting study of both cultures may be read in the authoritative book, "The Khoisan Peoples of South Africa", by Professor I. Schapera.

The Khoisan languages are full of fascinating clicks and many syllables. The "K" sounds in Okiep, Koukamma, Kamiesberg and Kamieskroon are all derived from their click consonants. The last two names incorporate the old Nama word, "kameeras", which means "bundle mountain" and refers to the rocky crown on the peak that overlooks the little town of Kamieskroon. Other names of Nama origin include Nababeep, Garies, Gamoep and Gariep. They are pronounced with a hard "g" and not with the guttural "g" used in Dutch and Afrikaans. Gariep is the old Nama name for the Orange River and it has been immortalised in the names of plants such as *Aloe gariepensis*, meaning the Aloe from the Orange River Valley.

The Hottentots are allied to the Bushmen, although there are striking differences in their religion and they did not paint on rocks as did the Bushmen. There are great resemblances in their language and appearance, and they were hunters like the Bushmen, but were also a pastoral people with herds of long-horned cattle and flocks of fat-tailed, hairy sheep. Like the Bushmen, they used plants to enable them to live off the arid land in which they lived and hunted, by eating some of them and using some as medicines, as well as in their traditional ceremonies.

Their arrows were carried in quivers made from the hollowed-out branches of *Aloe dichotoma*, in combination with antelope hide. These arrows, tipped with triangular iron pieces fitted on to a short bone-shaft and then into a main shaft of reed, were often tipped with poison. This was derived from the virulent milky latex of *Euphorbia virosa* that is found in Namaqualand, as well as from other plants and venomous snakes. They greased their arrows and quivers with the latex of another species, *Euphorbia stellaespina*.

Bulbs, which store water, were used as food by the Khoisan people, as were edible roots and berries. The sama Melon (*Citrullus lanatus*) is renowned for its use as food and moisture. Bulbs and corms of *Babiana*, *Gladiolus* and *Lachenalia hirta*, as well as the sweet watery root of *Cyphia crenata*, were eaten. They chewed the juicy stems and roots of *Albuca altissima* to quench their thirst and ate the roots of a species of *Solanum* and *Pelargonium fulgidum*, as well as other species of *Pelargonium*. A brew of wild honey was fermented with the help of fleshy stems like those of *Pachypodium*.

The name of "Kougoed", meaning "something to chew", was given by the Dutch to the plant called "Kanna" by the Khoisan people, who chewed it almost all day on account of "its pleasant smell and stimulating taste", as reported during the time of van der Stel's expedition to Namaqualand in 1685. This is *Sceletium tortuosum*, formerly called *Mesembryanthemum tortuosum*, which is now known to contain the narcotic "*mesembrine*." It was used medicinally in the relief of toothache or stomach-ache and as a substitute for tobacco.

The rough fleshy leaves of *Mesembryanthemum barklyi* were used to remove hair from hides, much like sandpaper. The waxy Bushman's Candle (*Sarcocaulon*) was eaten and also used as a fuel.

Sarcocaulon was also used medicinally as a remedy for diarrhoea and the powdered root used as a poultice. The bulb of *Veltheimia capensis* was used as a purgative, as were the berries of *Chironia baccifera*, which were also made into a decoction for boils. *Berkheya* was powdered to make a plaster for boils, while both *Berkheya* and *Passerina* were used to ease shooting pains. Pains in the back were treated with the bark of *Boscia* and Wild Rosemary, *Eriocephalus africanus*, was used for stomach-ache. The

PLATE 13

1 Fascinating bi-coloured flowers of *Nemesia versicolor* may also be pink or blue. This is common in Namaqualand and the S. W. Cape.

2 *Nemesia macroceras* is a slender annual that grows in sandy soil in the shade of small bushes.

Lachenalias are common in Namaqualand and, despite their small size, very attractive little bulbs. They are best known as Viooltjies.

3 Two fleshy oval leaves distinguish this richly-coloured robust species, *Lachenalia ovatifolia*, which grows near Springbok.

4 *Lachenalia violacea* may be identified by its single leaf and bluish-green flowers which are tipped with purple.

1

3

2

4

1

2

PLATE 14

1 The satiny flowers of
Grielum humifusum
shimmer in the sun-
shine. It has various
common names such as
Platdoring (Flat-thorn),
Duikerblom (Duiker-
flower) or Pietsnot.
Whether alone in great
sheets or in combination
with other flowers, it is
always lovely.

2 A swathe of *Grielum*
runs between orange
Ursinias.

PLATE 15 →

1 *Grielum humifusum* lies
in clumps between blue
Wahlenbergia, mauve
Felicia and yellow
Gazania.

2 It gleams here on the
red sands beside yellow
Senecio, Cotula leptalea
and mauve *Felicia
namaquana*.

1

2

leaves of *Salvia africana-lutea* were used to treat colds and fevers, as were the leaves of some Crassulas. The stems of some *Cotyledon* species were pounded and used as a poultice. The red flowers of *Sutherlandia frutescens* were made into a liquid that was used for fevers or for washing wounds, while the red flowers of *Pelargonium fulgidum* were used to combat weakness and anaemia. *Carpobrotus* species were used in child-birth, as were the boiled leaved of *Rhus* species.

THE EARLY SETTLERS

As early as 1660, Jan van Riebeeck sent an expedition into Namaqualand in order to search for the legendary kingdom of Monomotapa, but treasure was only found when Simon van der Stel led an expedition into Namaqualand in 1685 in order to find copper. His first shaft in the Copper Mountains, the Koperberg, is preserved as a National Monument and may be visited outside the town of Springbok, off the road leading to the east.

The daily Journal that was kept of this expedition, as edited from the manuscript by Gilbert Waterhouse, makes fascinating reading. The large travelling company set off from the Castle on the 25th August, 1685 and consisted of 57 men and their servants, Hottentot interpreters, 7 waggons, one loaded with a boat, 200 extra oxen, horses, mules and an escort of burghers and their waggons, which was to accompany them for part of the way. They had some remarkable experiences, being charged by a rhinoceros and having to head off elephants by creating a din with bugles and drums. Commander van der Stel's clever leadership won the friendship of Hottentot and Bushmen tribes, who helped to guide him to his goal. He gave them presents of tobacco, mutton, rice and brandy, entertained them and showed them kindness, even helping them in their wars and treaties. His journal is full of observations on their manners and customs, as well as references to their uses of plants. There is an engrossing account of the countryside, its birds and animals. The travellers, in much the same way as those of today, were stung by a yellow and black "blind fly" that "only disappeared when a kind of flower, very like the marigold, begins to wither".

After finding copper, the company went towards the sea, as far as the mouth of the river which they named the Buffels, after the Sonquas (Bushmen) had told them that they called it Tousé, meaning Buffaloes. An exploratory party went northwards and discovered the mouth of the Orange River, which they called the Eyn, and pronounced it "unfit for any ship to approach" on account of its violent seas and numerous rocks. Turning homewards, they arrived back safely on the 26th January, 1686. During the journey, the artist Claudius, who was employed by the Dutch East India Company, made drawings of the people, places, animals, birds and flowers that they encountered along the way. These may be seen at the Africana Museum, Johannesburg, as well as in the Catalogue of the Pictures in the Africana Museum (Vo. 2), compiled by J. F. Kennedy. They are also reproduced in the book on the Journal.

PLATE 16

Only in Namaqualand could one find a "middelmannetjie" sprinkled with a large variety of flowers, including both annuals and bulbs.

By 1780, Dutch farmers had settled in the Kamiesberg Range, where water supplies were comparatively good. In 1777, Captain Gordon, Commander of the Garrison at the Cape of Good Hope, crossed the river which bore the Nama name of Gariep, and renamed it the Orange River, in honour of the Prince of Orange.

Missionaries came to Namaqualand from the beginning of the 19th century, starting with the London Mission Society. The north-western region, which lies within the bend of the Orange River, was named the Richtersveld by the Rhenish Mission Society, in honour of the Rev. W. Richter, Inspector of the Society's Seminary in Germany. In 1839, the Dutch Reformed Church established a new congregation to serve the growing community of Namaqualand.

The place-names of Namaqualand reflect the Dutch language and interests of the early settlers. The once teeming wild life of the country can be imagined with names like Gemsbok Vlei, Elands Klip, Ratel Kraal, Quaggafontein, Steenbok, Springbok, Aardvark and Hartebeest Vlei. Plants are featured in some names like Bloemhoek, Swart Uintjies River, Kameelboom Vlei, Taaibosch Vlakte and Leliefontein. Some names, like Keerom and Beenbreek, mark the hardship of their pioneer lives, while places like Stinkfontein are not without humour.

The Dutch farmers of the 18th and 19th centuries were interested in the flowers they saw in Namaqualand, not only because of their beauty, but because of their medicinal qualities. In many cases, they were introduced to the uses of plants by the Nama people, but they were very much aware of herbs, which played a great part in their lives in the far-off settlements, when the women had to deal with ailments without the help of doctors. The common names of many plants bore testimony to some of their uses. Originally spelt in the Dutch manner, modern spelling has now brought these into line with the Afrikaans language. Wherever possible, English translations have been given in parenthesis in this text.

Droedas Kruiden, meaning "healing herb", was the name given to *Pharnaceum lanatum*. Katkruid, meaning "Cat-herb", was the name for *Ballota africana*, which was used by the early colonists as an infusion for colds and was taken by the Nama people to relieve colic.

Sometimes they named the flowers after the garden flowers they knew from Europe, such as "Tulp", meaning "Tulip", for *Homeria* or *Moraea*, or "Viooltjie", meaning "Violet", for the purple *Lachenalia*. Most of the common names given to flowers, however, were descriptive. "Spinnekopblom", meaning "Spider Flower", was the name given to *Ferraria; Ixia* was called "Kalossie", meaning "Little Bell"; "Kannetjie", meaning "Little Cans", was the name for *Microloma;* "Leeubekkie", meaning "Lion's Mouth", for *Nemesia;* "Veldskoenblaar", meaning "Leaves like a home-made shoe", for *Massonia latifolia* and "Fluitjies", meaning "Little Flutes" for *Lebeckia*, with its reedy stems.

The descriptive names are not always reliable in knowing which flower is referred to, as the same name was frequently given to different plants. "Gousblom", meaning "Gold Flower", was given to almost any gold or orange daisy. "Aandblom", meaning "Evening Flower", was given equally to *Gladiolus arcuatus*, which is fragrant at night, and *Hesperantha*, which opens towards evening. "Renosterbos" is the name given to both *Pteronia* and *Elytropappus*. There are several plants called "Katstert", meaning

"Cat's Tail", which include *Bulbine*, *Bulbinella* and *Struthiola leptantha*, so that it is important to know the botanical name in order to be sure of the plant to which one refers.

The flowers of Namaqualand are not so familiar or well-known that they were all given common names, however, and where a particularly interesting flower has not acquired a known common name, I must confess to have invented a few, such as Beetle Daisy for *Gorteria diffusa* subsp.*diffusa*, Springbok Painted Petals for *Lapeirousia silenoides* and Desert Broom for *Sisyndite spartea*. Wherever possible, however, one should try to learn the botanical names and use them constantly.

THE EARLY BOTANISTS

The plant collectors of the 18th and 19th centuries concentrated their findings at the Cape of Good Hope where the rich flora enthralled them. The more adventurous collected plants in the hinterland, but very few went as far as Little Namaqualand, for they risked dying of thirst, amongst other hazards. The names of many South African and European botanists, however, are commemorated in the names of Namaqualand's plants, even though they did not visit this region.

Paul Hermann, the distinguished Dutch botanist who visited the Cape in 1672, is remembered in the genus *Hermannia*, which has several species in Namaqualand, notably the lovely Desert Rose, *Hermannia stricta*. Francis Masson, the first botanist to be sent overseas by Kew, arrived at the Cape in the ship taking Captain Cook on his second voyage around the world in 1772 and made several journeys into the interior in order to collect plants. Although he did not penetrate as far inland as Namaqualand, his name is preserved in the Namaqualand bulb, *Massonia latifolia*.

The drawings of plants at the Cape by Hendrik Claudius in the 1680's, are of the greatest importance in the history of South African botany. He accompanied Simon van der Stel's expedition into Namaqualand and made drawings of many species. These included bulbous plants like *Albuca altissima*, *Babiana*, *Gladiolus*, *Lachenalia hirta*, *Lapeirousia* and *Veltheimia*; annuals like *Nemesia*, which he called "a kind of pansy", *Heliophila*, which he named *Linaria*, and *Gorteria*, which he thought to be a *Gazania*. Other drawings of the plants seen in Namaqualand include several kinds of *Pelargonium;* trees and shrubs like *Acacia karroo*, *Ozoroa*, *Diospyros* and *Lebeckia;* succulents like *Sarcocaulon*, *Cotyledon*, *Sceletium tortuosum* and several kinds of *Aloe* and *Euphorbia*.

During these early days in Cape history, Commander Robert Jacob Gordon was in charge of the garrison at the Cape of Good Hope. He travelled through Namaqualand in 1777 and named the Orange River after the Prince of Orange. The name of this cultured Dutch soldier, whose numerous drawings of plants are in the Rijks Museum in Amsterdam, lives on in the succulent called *Hoodia gordonii*. The name of Sir Henry Barkly, who was Governor of the Cape of Good Hope from 1770 to 1777 and an ardent naturalist who was interested in succulents, is remembered in the succulent plant, *Mesembryanthemum barklyi*.

Johann Franz Drège, the outstanding and indefatigable German professional plant collector, who spent eight years at the Cape, went on a lengthy expedition to Namaqualand in 1830. His vast collections are found in most of the world's largest herbaria and several plants from Namaqualand bear his name, including *Babiana dregei* and *Euphorbia dregeana*.

Carl Philip Zeyher, the German plant collector, went twice to Namaqualand, in 1829 and 1830. His name is remembered in *Cotula zeyheri* and *Salsola zeyheri*. Sir James Edward Alexander was a geographical explorer who went to Namaqualand about 1840. He crossed the Orange River and collected plants on his travels, about which he wrote two volumes.

Several people who were active in Namaqualand in the second part of the 19th century had plants named after them. Albert von Schlicht, a German pharmacist by training, settled in Concordia in the late 1850's, where he prospected for copper. His plant collections were sent to Hamburg and he is remembered in the succulent *Psilocaulon schlichtianum* (formerly *Mesembryanthemum*). The bulb *Whiteheadia bifolia* keeps alive the name of the Rev. Henry Whitehead, a Church of England clergyman, who collected plants when stationed at the copper mines in Namaqualand in the late 1850's. W. C. Scully, a writer who was a border magistrate for Namaqualand in the late 1850's, is honoured in the names of *Homeria scullyi*, *Gladiolus scullyi* and *Spiloxene scullyi*, which are found in Namaqualand. *Homeria schlechteri* commemmorate Rudolf Schlechter, a professional plant collector in Namaqualand in 1897.

The turn of the 20th century marked an ever-increasing interest in the flora of Namaqualand, which grew as travel became easier. Professor W. Harold Pearson, founder of Kirstenbosch Botanic Garden, went to Namaqualand in 1903 and had *Aloe pearsonii* and *Lobostemon pearsonii* named after him. The famous Dr. Rudolf Marloth, who travelled far and wide in South Africa, collecting plants until his death in 1931, visited Namaqualand on his way to South West Africa. *Homeria herrei* was named after Hans Herre, the writer on *Mesembryanthemaceae*, who collected succulents extensively in Namaqualand from 1929 onwards.

Botanical explorations by numerous botanists and plant enthusiasts have continued in Namaqualand to the present day. In this little-trodden region it is still possible to roam and experience the excitement of the unknown and the thrill of seeing sights that few, if any, have seen, as well as to discover plants that have been growing and flowering untended, since the beginning of time.

THE ANNUALS
AND
PERENNIALS

The Annuals and Perennials

Annuals of all kinds seed themselves in profligate splendour in Namaqualand each season, waiting for the autumn rains to bring them to life. They seldom need to wait for longer than a year in order to complete their cycle of life and perpetuate their kind. They do not all germinate each year, otherwise there would be a danger of the species dying out in a bad year. For this reason some years are better than others. This regeneration has been taking place continuously over the centuries, and they will continue to propagate themselves for as long as man will leave them untouched by the plough or by progress. It is a comfort to imagine that the mining operations in the mineral-rich soil of Namaqualand could continue beneath the earth's crust, so that the evanescent yet priceless treasures of the wild flowers will be preserved for the pleasure of man in the future.

When one thinks of the flower spectacles in Namaqualand, one thinks chiefly of daisies and it is true that these sweet and simple flowers do most towards creating the sheets of flowing colour that are so breathtaking in their extravagance and intensity. The immediate picture that they call to mind is that of a blanket of brilliant colour lying at the foot of one of the typical conical hills in the area, curling softly in wavelets up against its contours and enfolding the boulders at the base. A sea of orange *Ursinia* might melt into a swathe of yellow *Cotula* and then, as it spills at one's feet near the roadside, may be broken into rainbow spots that emerge as small treasures hidden between the daisies. Sometimes one can distinguish tiny bulbs like *Oxalis* flowering between the daisies, forming a second layer beneath them, while, at other times, chunky mounds of thick-leaved succulents burst from between them. These may be silky yellow *Conicosias* or shimmering purple *Lampranthus*, reflecting light like a mirror on a sunny day.

At other times, one can come upon a living Persian carpet in a low valley, composed of millions of annuals in a kaleidoscope of colour, embroidered by bulbs and succulents of many kinds. In her colour combinations, nature often proves to be the greatest artist of all, harmonizing drifts of plants like clear mauve *Senecio* with patches of lemon-coloured Button Flowers (*Cotula barbata*).

Other combinations of annuals seen frequently include Orange Namaqualand Daisy (*Dimorphotheca sinuata*) with dark blue *Felicia*, coppery-orange *Gorteria* or azure blue Sporries (*Heliophila*), or in a mixture with white Sporries (*Heliophila*) and yellow *Senecio* or pale blue *Felicia*. Dainty white and blue Sporries are often seen together, as well as alone in great masses. Light orange *Osteospermum* seems to spring from the crevices in rocks and may be seen in combination with yellow *Senecio* and white Sporries. Pale blue *Felicia* and dwarf yellow *Cotula* make an excellent combination and the blue Felicia also offsets the delicate chartreuse-yellow of silky *Grielum*.

"Gousblom" is the Afrikaans word meaning a golden flower, and this is a name that has been loosely used to describe any daisy with a yellow or orange flower, even though it may belong to differing botanical types, such as *Arctotis*, *Dimorphotheca*, *Gazania* and *Ursinia*. Some have thin petals while others are broad; some are single and some double; some dwarf and some tall, but they all have differing characters or leaves that enable the botanist to distinguish between them, although this is not always an easy task, even for the taxonomist.

Some annuals seed themselves so freely that they lodge in thatched roofs which are then coated with livery in the spring. Miniature yellow Button Flowers (*Cotula barbata*) are seen most frequently on farmhouse roofs, for they grow with ease and abandon. On sandy country roads, where wheels have worn a double track, leaving a humped strip or "middelmannetjie" down the middle, apricot-coloured Gazanias or Namaqualand Daisies run along the central ribbon like a flame. On the borders of green wheat fields, small daisies like those of *Gorteria* form a fringe of coppery orange. Wherever there is a foothold for these annuals in the more favourable areas, they overflow and bloom in exuberant abundance.

Perennial plants eke out their existence from one season to the next, enduring the searing heat and drought of summer until they are revived by soft winter rains, so that they will bloom in springtime. They can be seen in masses at the roadside, where most moisture gathers in the depressions on the verges. Brilliant saffron and tangerine Gazanias spread on the banks in dazzling clusters; small Pelargoniums, with striking magenta flowers, stand out like signals among the golden daisies, while cushions of sapphire Karoo Violets (*Aptosimum*) stud the red soil of the driest areas like rich jewels.

Both the annuals and perennials that grow wild in Namaqualand, more especially those that belong to the daisy family, open fully only on sunny days and then only during the warmest part of the day, between ten o'clock in the morning and about four in the afternoon, so that the enthusiastic traveller should plan to be amongst the flowers during the warm and sunny period and take a picnic lunch in order to make the most of the midday hours.

Many of the annuals and perennials listed below are the most commonly seen in the flower fields, but some of the beautiful or interesting, though rarer, kinds are also mentioned, in case the amateur botanist would like to identify them. The descriptions are brief of necessity, but if in doubt, the botanical names can be checked in botanical books, as listed in the bibliography at the back of the book, or at an Herbarium.

PLATE 17

1 In a pastoral scene near Springbok, sheep wander into the low bushes that cover the hillsides.

2 In the Hester Malan Wild Flower Reserve near Springbok.

3 Boulders provide shade for picnickers and plants in Namaqualand.

PLATES 18/19

Pastel-shaded *Arctotis canescens* spread over the vast sandy plain between Pofadder and Aggeneys, on the way to Pella. The large daisies may be pale orange, salmon or white, with many shades between, for they hybridise in nature.

1

Amellus hispidus Amellus
(Compositae)

A perennial with daisy-like saxe blue flowers in early summer, this has a broad yellow centre about 2 cm in diameter, while the narrow blue rays that surround it are each 1½ cm long. It has branching stems growing to 35 cm, that bear rough, narrow leaves, each about 3 cm in length.

A. *hispidus* occurs in damp spots near Leliefontein in the Kamiesberg and is found in many places from Kamieskroon to Springbok. A smaller form occurs in the Richtersveld and a form with smaller flowers and narrower leaves is found in the S. W. Cape.

Anchusa capensis Cape forget-me-not
(Boraginaceae)

This well-known annual grows wild all over the Cape, O.F.S. and Lesotho, as well as in Namaqualand, where it may be seen in the moister parts of the mountains, near Leliefontein in the Kamiesberg and in the southern area. It flowers in springtime, bearing its deep blue Forget-me-not flowers at the top of a branching stem. This grows to about 45 cm and the rough, hairy leaves form a tuft at the base of the plant.

Aptosimum Karoo Violet
(Scrophulariaceae)

These low-growing perennials owe their beauty to the rich royal-blue or violet of the tiny, funnel-shaped flowers, each about 2 cm long, with white markings at the throat. Despite their small size, they are so crowded on the plants that they command attention. Four species occur in Namaqualand and the first two differ slightly according to their leaves.

A. *depressum*. Probably the most attractive, as the flowers are so closely massed above the leaves, this has exposed flowers that form a brilliant little mat, spreading about 25 cm across. The soft, oval green leaves are distinctly pointed. This species grows in hot dry areas west of Pofadder on the road to Springbok (see plate 21).

A. *indivisum*. This little perennial also forms a flat mat on the ground, but the flowers, similar to those of A. *depressum*, are half obscured by the crowded, spathulate leaves, as their stalks or petioles are nearly as long as the leaves themselves.

A. *leucorrhizum* (*Peliostomum leucorrhizum*). A delicate bush annual that grows to a height of 30 cm, this has deep violet flowers with a long, dark, lined throat opening into

PLATE 20

1 Wherever a small stream trickles down from rocky kloofs in the hills, moisture-loving flowers collect, seeding themselves on narrow ledges of damp soil to form little gardens of interest and beauty.

2 The delightful little purple *Monopsis campanulata*, found only on damp sand, is seen here as a companion to the Cape Weed, *Arctotheca calendula*, which has many forms and grows in a variety of places.

5 petals. They are scattered along the thin branches that are covered with very short, needle-like soft leaves. This plant grows in association with *A. depressum* west of Pofadder.

    A. spinescens. This branched plant is quite distinctive, as it has needle-like, spine-tipped leaves that are densely crowded along the stems. The flowers are thickly scattered along the branches between the leaves (see plate 22).

Cape Weed   *Arctotheca calendula*
(Compositae)

This low-growing perennial is common in the Cape and Namaqualand, where it spreads over huge fields, producing masses of yellow flowers in springtime. They are about 4-5 cm in diameter and may be bright yellow or lemon yellow. The backs of the petals are normally greenish or marked with thin grey lines. The leaves are hairy and greyish green, but some forms have green leaves backed with silvery felt. There are many forms, some of which behave as annuals (see plates 20, 33).

Arctotis, Gousblom   *Arctotis*
(Compositae)

Several species of this large, attractive genus grow wild in Namaqualand, often forming great colonies of flowers for miles on end. Some have large flowers, with brilliant colouring, and create an unforgettable sight of stunning beauty when they open fully in the midday sunshine. They vary considerably in height, according to the amount of moisture received during the season, but when there have been good rains, people drive for hundreds of miles to marvel at the spectacle.

    A. canescens. Often referred to by Springbok residents as "pastel-shaded" *Arctotis*, these annuals may be seen east of Springbok, on the vast sandy plain between this little town and Pofadder. Sometimes they spread as far as the eye can see, while, at other times, they are sprinkled in groups between other annuals. They form bushy plants about 55 cm high, completely covered with scores of large pale orange or yellow flowers, each fully 5 cm in diameter. There are pure white forms and numerous salmon shades, for they hybridise in nature. The divided leaves are felted on both sides, so that they have a corase greyish-green appearance (see plates 18, 19).

    A. crispata, from the Kamieskroon area, has very narrow leaves and smaller flowers, about 3,5 cm wide.

    A. fastuosa (Venidium fastuosum). Familiarly called the Double Namaqualand Daisy, this has large brilliant orange flowers, from 7 to 10 cm wide, with ray flowers or "petals" that stand up alternatively and seem to form a second row. Gleaming black patches at the base form a ring around the broad purplish-black, shining centre. These flowers are borne at the tops of tall stems, growing to 60 cm, and the shaggy leaves are grouped mainly near the base. The leaves and stems are covered with a whitish felt, formed by cobwebby hairs. This is also known as the Bitter Gousblom as it imparts a bitter flavour to milk when it is grazed (see plates 5, 40).

This species grows in great tracts around Springbok, extending to Steinkopf in the north and on into S.W. Africa. This is the most brilliant of the *Arctotis* species in Namaqualand and has long been a favourite in gardens.

A. hirsuta is similar to *A. fastuosa*, but has small orange flowers ringed with black. It is wide-spread in Namaqualand, the Karoo and S.W. Africa, extending southwards to Van Rhynsdorp.

A. leiocarpa, the Karoo Daisy, looks like a white form of *A. fastuosa*, generally with a "double" row of shimmering white petals. The centre is yellow, often with a central black spot. This spring-flowering annual grows to 55 cm and the greyish-green stems and divided leaves are covered with silvery hairs. This species grows wild in Namaqualand, west of Springbok, as well as in the Karoo, north-western Cape and S.W. Africa.

A. laevis (*A. squarrosa*) is a shrubby plant with branched stems and finely divided dark green leaves, which are crisped and curly. The large orange flowers have orange centres and are borne on long stems above the foliage, flowering profusely in spring. This species is seen on rocky hillsides in the S.W. Cape, near Clanwilliam, and extends northwards into Namaqualand.

Ballota africana   Cat-herb, Katkruid
(Labiatae)

Only one species of this large drought-resistant genus occurs in South Africa, spreading along the south-western coast from Namaqualand to the Eastern Cape Province. It has small two-lipped, pale pink, mauve or purple flowers with 3 lobes on the protruding lower lip. These are arranged in few-flowered whorls on the grey-woolly branches. The toothed leaves are larger near the base of this small shrubby perennial, which is not very showy. This was used medicinally by the Nama people and the early colonists.

Castalis tragus   Castalis
(Compositae)

This small perennial has an orange daisy flower that is about 2 cm across, borne singly at the top of the stem. Several stems form a tufted plant. The leaves, about 4 cm in length, are simple and barely indented. A woody rootstock enables it to resist drought. Specimens of the plant have been found near Okiep and it also occurs in the S.W. Cape.

Charieis   Charieis
(Compositae)

The True-Blue Daisy (*Charieis heterophylla*), with its azure blue petals (ray flowers) and deep blue centre (disc flowers), is well-known to discriminating gardeners and the blue centre is characteristic of this species. This annual grows wild in the S. W. Cape and was described in Flora Capensis by the botanist Harvey, who stated that there was both a blue-centred and yellow-centred form of *C.heterophylla*.

Recent investigation at the National Herbarium in Pretoria, however, has led to the belief that this yellow-centred form is a new species, as yet undescribed. This was collected near Springbok by the author and is illustrated in this book. It is referred to as *Charieis* sp.nov (see plate 6).

It is impossible to distinguish with the naked eye between this plant and *Felicia bergeriana*, as one must use a magnifying glass in order to observe the botanical feature that differentiates them. In *Felicia*, each small ray floret always has a *pappus* or tuft of downy hairs on the ovary, while the ray florets of *Charieis* are always smooth.

The unnamed secies of *Charieis*, as illustrated, has rich, dark cobalt blue flowers, measuring about 4 cm across, with a yellow centre. The ray florets seem to be narrower than those of *Felicia bergeriana* and the plant appears to grow taller, up to 20 cm in height. The slightly hairy leaves are 3-4 cm long and ½ cm in width. The only way to differentiate between the two, however, is by noting the absence or presence of *pappus* on the ovaries of the ray florets. Both plants are annuals that bloom in spring and may be found in the fields around Springbok.

Chascanum — *Chascanum gariepense*
(Verbenaceae)

A small undershrub, this has white flowers on a long spike, opening upwards to the top. Each flower has a 2 cm long narrow tube, opening into 5 lobes, and is reminiscent of *Sutera*. The fairly small dentate leaves are about 3 cm in length and 1½ cm wide. This species grows in the northern Cape and Namaqualand, near Upington, Goodhouse and Springbok, as well as in S.W. Africa.

Grey-leaved Cineraria — *Cineraria canescens*
(Compositae)

This tall annual grows to 5 cm and has massed heads of small yellow daisy-flowers, each about 1 cm long. The greyish-green leaves are broad and rounded with toothed and indented edges. They sometimes grow to 5 cm in length, but are often much smaller.

This plant is found in the shade of rocks and in sandy places near Springbok, as well as in S.W. Africa and the Transvaal.

Tsama Melon, Wild Water-melon — *Citrullus lanatus (C. vulgaris)*
(Cucurbitaceae)

The smooth, large, spherical fruits of this prostrate annual are well-known as a source of food and moisture for the Bushmen. They grow up to 20 cm in diameter and are pale greenish-yellow, usually mottled with bands of dark green. The saucer-shaped flowers are about 3 cm in diameter. The large, rough, oval or triangular leaves are deeply lobed and may be from 6 to 20 cm long and 4 to 15 cm broad. They are attached to long stems, growing to 3 metres, which have tendrils and sprawl on the ground. This plant grows in the Richtersveld.

Cotula
(Compositae)
Button Flowers, Gansogies, Duck's Eyes

These tiny flowers resemble daisy centres, but they grow in such enormous drifts that they colour whole fields golden and seed themselves profusely everywhere, even in the thatched roofs of farm houses.

C. barbata. This is the most common species, with numerous yellow pea-sized flowers that are borne at the tips of erect, wiry stems growing to about 10 cm. They obscure the small feathery greyish-green leaves that form a tuft on the soil. This species has an unpleasant odour, giving rise to the common name of Stinkkruid (Smelly Weed) (see plate 7).

C. leptalea has much longer stems that bear the yellow flowers about 10 cm above a neat cushion of greyish-green, feathery foliage. The flowers are not so numerous or dense as those of *C. barbata*, but are larger, about 1$\frac{1}{2}$ cm across. This species is common around Springbok and Kamieskroon (see plate 15).

H. laxa is similar to *C. leptalea*, but the leaves are much divided. It has a large, long-stemmed flower and grows wild in the Kamiesberg mountains.

C. thunbergii is a small plant with short, finely divided leaves. The stems are very long, up to 25 cm, and the yellow flowers are 1$\frac{1}{2}$ cm in diameter.

C. zeyheri is a dwarf plant with very tiny white button-flowers and may be seen near stream-sides in the Springbok area, but it is not common.

Cyphia crenata
(Campanulaceae)
Veld Barroe, Aard Boontjie

This dainty climbing perennial twines into low bushes, flowering in springtime and dying down during the dry summer. The small white flowers have 5 narrow pointed petals joined into a short tube, and are usually pink on the reverse side. The plant has a tuberous root. It grows wild in hilly places among bushes in Namaqualand. Another similar species, *C. volubilis*, differs from it in small technical details.

Cysticapnos africana (C. vesicarius)
(Fumariaceae)
Klappertjies, (Little Rattles)

A delicate climbing annual with very soft, sappy stems, this grows to about 50 cm and has tiny, soft tendrils that enable it to climb into low bushes around it. The small, pale pinkish-mauve flowers, often spurred, appear in spring, followed by inflated veined seed-pods that give rise to the common name. This genus has only one species which is found growing in sandy soil in the mountains in Namaqualand, as well as southwards into the Cape Peninsula.

Dianthus scaber
(Caryophyllaceae)
Wild Pink

So small and delicate as to be almost inconspicuous, this little perennial grows to about 10 cm in height and bears small whitish single flowers with serrated edges at the top of erect stems. It is found in sandy soil among other plants in the dry areas east of Springbok, flowering in springtime.

Twin-spur, Horingtjies	*Diascia*

Diascia
(Scrophulariaceae)

There are several attractive species of this large S. African genus in Namaqualand, with conspicuous reddish-purple flowers. These are two-lipped and have rounded lobes, while the lower portion generally has two pouches or spurs. They are difficult to distinguish from one another, but all are attractive. They are often found growing in sandy, damp soil near river-beds, but also occur in drier places, flowering in springtime.

D. floribunda is a short plant with small leaves and large, reddish-purple, twin-spurred flowers.

D. namaquensis has small purple flowers with two short spurs.

D. nemophiloides. An erect, thin-stemmed plant, this has deep purple flowers about 1½ cm across. It is common in the Cape and may be seen west of Springbok near the foot of Spektakel Pass and near Kamaggas.

D. sacculata is a low plant with very tiny reddish-purple flowers, scarcely ½ cm across, and longish leaves.

D. tanyseras. Very similar to *D. floribunda*, this has dark mauve flowers with 2 long spurs about 2 cm in length.

D. thunbergiana. This tall annual grows to a height of about 25 cm, with a tuft of soft indented leaves at the base. It has deep purplish-mauve or lilac flowers with 2 long conspicuous "horns" (about 1½ cm long) that have evoked the local name of "horingtjies," "little horns". It is common around Springbok and Kamieskroon (see plate 27).

Namaqualand Daisy, African Daisy, Rain Daisy, Botterblom

Dimorphotheca
(Compositae)

Two of these species are among the most attractive of all the daisies in Namaqualand, while *D. sinuata* is so spectacular and common, painting huge tracts of land brilliant orange, that it has become known as the Namaqualand Daisy. It is also known as the African Daisy in gardens throughout the word, where it is popular. The colour forms found in cultivation are believed to have developed through selection.

D. sinuata (*D. aurantiaca*) is an annual that forms a bushy plant, often growing to 40 cm in height, that is covered with large flowers, all blooming at the same level and forming a glistening mass that is most brilliant when the flowers open fully in bright sunshine, from mid-morning to mid-afternoon. Each flower is about 8 cm in diameter and composed of a single row of oblong florets that are greenish-purple at the base, forming a ring around an orange centre. This is often referred to as the Single Namaqualand Daisy, as distinct from the so-called Double Namaqualand Daisy (*Arctotis fastuosa*). The indented narrow leaves are larger near the base of the plant. *D. sinuata* is common in the fields around Springbok and Kamieskroon, blooming in combination with other spring annuals (see plates 1, 2, 9).

D. pluvialis, known as the RAIN DAISY, CAPE DAISY, or WITBOTTERBLOM, has a glistening white flower, with its petals backed with mauve, and a yellow centre. A deep violet ring around the centre varies in width and is sometimes scarcely apparent. This annual grows to about 30 cm in height and the flowers cover the bush in springtime, forming a dazzling mass. The Rain Daisy covers huge fields in the S.W. Cape and is also found in Namaqualand.

D. polyptera is another Namaqualand species that is not as showy as the others. It has small greyish-yellow flowers, about 1½ cm in diameter, and finely divided leaves.

Drosera Sun Dew, (Droseraceae) Flycatcher

A few of these fascinating little perennials occur in damp sandy places in Namaqualand, but these are so tiny and difficult to identify that no particular species can be mentioned here. Like all Sundews, they have sticky tentacles on the thin leaves to which small insects adhere. The 5-petalled flowers are satiny and vary in colour from white to pink, mauve and purple. Sundews have underground rhizomes that store moisture and bloom during spring.

Erodium botrys Erodium, (Geraniaceae) Turksnaald (Turks Needle) Heron's Bill

A tiny prostrate perennial, this has a flat rosette of small, narrow, deeply indented, simple leaves on long stalks. The flower stems grow up from between the leaves to a height of 15 cm, with small pinkish-mauve flowers at the tips. The long, pointed fruit suggests the common names. The plant has a woody taproot. It occurs at Grootvlei, west of Kamieskroon, in damp spots. There are large forms with longer trailing stems that come from the S.W. Cape.

E. cicutarium. This little perennial occurs near Leliefontein and Hondeklip Bay, as well as in the S.W. Cape. It may be recognised by its ferny foliage that consists of leaves, about 4 to 7 inches long, that are divided and subdivided into tiny leaflets. The flowers are light mauve.

Felicia (Aster) Felicia, (Compositae) Kingfisher Daisy

Azure blue flowers are the chief attraction of this genus, although several species have mauve or white flowers. They all have yellow centres. The Felicias that grow wild in Namaqualand often cover great fields with pale or deep blue, making an impact even at a distance. They combine in great drifts with golden Cotula or form a refreshing break amid orange Namaqualand Daisies or terracotta Gazanias.

F. bergeriana, the Kingfisher Daisy, is an extremely common annual in Namaqualand, as well as in the S.W. Cape. It has cobalt blue flowers, measuring about 4 cm across, with a yellow centre. The petals unfurl fully in the late morning and roll back to close in the late afternoon. The leaf is narrow and hairy, about 3 – 4 cm long and 3 mm wide. The Kingfisher Daisy grows to about 20 cm in height, spreading over the ground, but may be much dwarfer in drier places. (See under *Charieis* for an annual which may be confused with this species).

F. adfinis. This annual is much more dwarf than the Kingfisher Daisy, usually about 10 cm in height. The flowers are pale blue rather than deep blue. The leaves are hairy, like those of *F. bergeriana*, but shorter and narrower. This species is common in Namaqualand and the S.W. Cape.

F. namaquana (*Aster namaquanus*). This bushy annual grows to about 30 cm in height, bearing large blue or mauve flowers, about 4 cm across, with a broad yellow centre. The weight of the long florets often makes them droop backwards slightly. The thin hairy leaves, about 3 cm long, may be up to 4 mm in width. The plants are individually dotted rather than in compact colonies. This species occurs throughout Namaqualand, the Karoo and S. W. Africa (see plates 6, 15).

F. scabrida (*Aster scabridus*). These sprawling perennials may be so dwarf through drought that they are prostrate on the ground, but form low, tangled bushes in more favourable conditions of growth. The pale mauve flowers are about 3 cm across and the very short, notched leaves have a rough texture. This species occurs in large drifts in dry river beds, such as in the Buffalo River west of Springbok, as well as near Concordia and Kamieskroon (see plate 45).

F. tenella (*Aster tenellus*). A low annual that may be prostrate on the ground or grow to 15 cm in height, this has pale blue flowers about 2 cm across. The flowers are larger in Namaqualand than those forms in the S.W. Cape. The flowers may also be cream or pale mauve. This species may be recognised by its soft needle-shaped, pointed leaves that vary considerably in length (see plates 6, 40).

Grey-plant *Forskohlea candida*
(Urticaceae)

A perennial growing to about 60 cm, this has woolly grey flowers and small dark green leaves which are thickly felted with grey beneath. They are roughly triangular, tapering at the base and toothed at the top. This occurs near Vioolsdrift and Strinkfontein in Namaqualand, as well as Kakamas in the N. Cape.

Galium *Galium capense*
(Rubiaceae)

A soft straggling perennial, this has stems to 60 cm that support themselves on low bushes. The very small flowers have 4 narrow segments and are grouped in heads at the tips of the branches. The hair-thin leaves are about 1 cm long and grouped in whorls of 6 to 8 on the stems. This species occurs in the Kamiesberg.

Gazania *Gazania*
(Compositae)

Brilliant when they open in full sunshine, Gazanias are attractive whether large or small. They are perennial plants that form low cushions on the ground, blooming in springtime.

PLATE 21
Stunning in their intensity of colour, the little flowers of the Karoo Violet, *Aptosimum depressum,* make this one of the most beautiful plants of the semi-desert.

1

2 3

1

PLATE 22 →

1/2 The delicacy of the pale blue Bell-flower,
Wahlenbergia annularis, is most apparent
when it is silhouetted against the sky.

3 A low, spreading annual, this lovely little Bell-
flower, *Wahlenbergis prostrata,* has unusual
blue flowers with a starry white centre.

2

3

← PLATE 23

1 Tiny purple flowers decorate the prickly stems
of this Karoo Violet, *Aptosimum spinescens.*

2 The Wild Mallow, *Anisodontea triloba,* has
pretty flowers on an open-branched bush. It
grows wild in the Kamiesberg.

3 The delicate spikelets of Katstert (Cat's Tail),
Struthiola leptantha, emerge from shady
nooks in the hills among other small shrubs.

1

2

3

G. krebsiana (G. *pavonia*), the TERRACOTTA GAZANIA, is one of the loveliest of all daisies, with its large flower, fully 8 cm in diameter, and its broad, bright orange centre, about 2½ cm wide. Each single floret is bright orange with a mahogany-red stripe down the centre, becoming scarlet near the tip. At the base of each segment there is a tan-coloured patch containing a white spot and a deep golden yellow base. This forms a richly coloured ring around the centre, with the whole forming a truly brilliant picture. This is the best colour form, but the flowers vary in size and colouring. The greyish-green leaves are deeply divided (see plate 8).

G. leiopoda is a low-growing perennial with divided leaves and large, deep yellow flowers, 7 cm in diameter, with a black ring at the centre.

G. lichtensteinii is a very common species in Namaqualand, the Karoo and S.W. Africa. The rich yellow flowers are 5 cm across with black spots near the centre of the petals. The plant is very flat on the ground with stems to 8 cm long and very short leaves (see plates 8, 15).

Gorteria
(Compositae)

Gorteria, Beetle Daisy

A deep coppery-orange daisy, this can be distinguished from afar by its colour, which is much deeper than that of the orange *Ursinia* or *Dimorphotheca sinuata*. It is also a more dwarf plant, hugging the ground and spreading out from the centre. Each flower is small, barely 2 cm across, but they vary in size. *G. diffusa* subspecies *diffusa* has pointed petals, while the subspecies *calendulacea* has very rounded petals. There is a dark brown or black mark at the base of the petals and a yellow centre. The leaves are thin and hairy.

An extraordinary feature of this plant is the presence of 3 tiny petals at the centre, which are iridescent green in colour and resemble tiny beetles at first glance. It is thought that these are meant to resemble female beetles, so as to attract male beetles to pollinate the flowers, while some believe that this is a device to repel destructive beetles. This phenomenon has not yet been studied scientifically.

G. diffusa supspecies *diffusa* is very common on the hills and plains around Springbok and Kamieskroon (see plate 25).

Grielum humifusum
(Roseaceae)

Platdoring, Duikerblom

A dwarf annual, this has large primrose-yellow, buttercup-like flowers with a satiny texture, so that they shimmer in the sunshine. They remain open until late afternoon. The sprawling branchlets and finely divided leaves are spread out on the surface of the soil, bearing the flowers in masses above the foliage. They lie in huge drifts of pale chartreuse yellow, distinctive from afar as they form huge patches on slopes and

PLATE 24

1/2 It is incredible that such dainty flowers as those of *Manulea nervosa* should not wither in the dry heat of the semi-desert. Both blue and mauve colour forms are common in Namaqualand.

3 Fragile flowers of Wilde Rabas, *Monsonia umbellata,* sprawl at the roadsides in the arid region near Aggeneys, unbelievably fresh despite the heat and drought.

banks, curving in waves around the base of the hills. Each flower has 5 broad petals that form a cup fully 5 cm wide, with white patches near the centre. It is followed by a flat rounded seed-pod, with an upstanding thorn at the centre, which gives it the common name of PLATDORING (Flat-thorn). It is also called DUIKERBLOM (Duiker-flower) as these little buck eat the flowers, and is known by the unflattering name of Pietsnot.

Grielum humifusum is wide-spread in all parts of Namaqualand, especially around Springbok. It is often seen in combination with orange *Gorteria* or *Ursinia*, as well as blue *Heliophila* or *Felicia*. It grows in dry sand and it is amazing that such a delicate-looking flower can survive its arid surroundings. It is slightly woolly on the stem and upper side of the leaves. There are several species of *Grielum*, which are all dwarf annuals with a thick taproot that enables the plant to resist drought (see plates 14, 15, 54).

G. obtusifolium is a Namaqualand species that is a very woolly plant. Some consider this to be the same as *G. humifusum* or a variation of it.

| Mountain Hamimeris | *Hamimeris montana* |

(Scrophulariaceae)

A small hairy annual that grows to 30 cm, this has bright yellow to deep orange flowers which resemble those of *Nemesia*. They are 8 to 15 mm broad, with 2 short pouches behind, and dark spots on the upper lip. The toothed, slender leaves grow to 4 cm in length. This little plant is found from the Kamiesberg to the hills around Springbok.

| Slak-blom, Snail-flower | *Hebenstreitia sarcocarpa* |

(Scrophulariaceae)

A slender branching annual growing to 25 cm, this has slender tapering spikes of tiny white flowers with yellow centres. The long needle-shaped leaves are soft-textured. This plant grows in the shade of bushes on the dry lower slopes of Spektakel Pass.

H. dentata var. integrifolia. This small annual has spikes of cream flowers with an orange mark, growing erect up to about 25 cm.

There are numerous members of this genus spread throughout S. Africa, some shrubby or perennial with woody rootstocks. *H. parviflora* and *H. stenocarpa* are two Namaqualand species with very much smaller flowers.

| Everlastings, Straw-flowers | *Helichrysum* |

(Compositae)

The straw-like texture of the flowers in this genus give rise to the common name. They are generally adapted to resist drought and thrive in rocky well-drained slopes in the mountains. There are several species with small yellow or white flowers that occur in Namaqualand, chiefly in the Kamiesberg, but only 2 have large or showy flowers.

H. sesamoides. This is a slender perennial shrublet, growing to about 30 cm, with white or yellow flowers, almost 4 cm wide, at the top of the thin stems that are covered with pointed leaves. This is a common species in the S. W. Cape and it occurs in the Kamiesberg in Namaqualand.

H. stellatum has smaller pink and white flowers, about 1 cm wide, and is found near Hondeklip Bay.

These dainty annuals form a delightful feature in Namaqualand's springtime spectacle. Myriads of these incredibly delicate plants form such enormous drifts that they lie like snow on the mountains or paint clouds of blue on bare red sands. At other times they emerge between short bushes and annuals, forming an effective foil to orange Namaqualand Daisies or Ursinias. Despite their fragile appearance, they seem to be able to resist great heat and drought when they bloom.

There are said to be over 100 species in S. Africa, massed mainly in the S. W. Cape, and the differences between the species are confusing. The flowers have 4 petals, which may be round or oblong, and vary considerably in size. The most common colours seen are blue or white, but some fade to rose or mauve, or may be lilac or yellowish. The stems and soft, slender leaves vary in size and shape.

H. bulbostyla is an annual that may grow to 50 cm, with bright blue flowers. The petals are a little shorter than those of *H. coronopifolia*, from 6 to 8 mm long. They are in a loose spike at the tops of the stems and the buds are rounded. The narrow leaves are from 2 to 7 cm long and the lower leaves may have 3 to 7 lobes, being grouped mainly near the base of the plant. This species is found on sandy soil near Garies, Kamieskroon, Droedap and west of Springbok (see plates 11, 54).

H. coronopifolia (*H. longifolia*). The most common blue-flowered *Heliophila* in the S. W. Cape, this is also found in Namaqualand, especially around Springbok. It has pale to bright blue flowers with a white or pale greenish-yellow centre. The petals may be rounded or oblong, from 6 to 13 mm in length and 3 to 8 mm wide. These are arranged in spikes near the top of the stems and the buds are oblong. This annual may grow from 10 to 60 cm in height, being taller and more branched where conditions of growth are good. The slender leaves, from 5 to 16 cm long and sometimes up to 3 mm broad, may be simple or divided into 3 or up to 13 lobes (see plate 11).

H. cornuta is interesting because it is a small shrub with woody stems and grows to 1½ metres in height. It has numerous tiny flowers that may be white or mauve. It occurs near Springbok.

H. namaquana. A slender plant with a main stem divided into 3 or 4 stems, this bears clusters of small white flowers at the top, sometimes fading to mauve or blue. The needle-shaped, simple leaves are soft, fleshy and up to 4 or 5 cm long. This is often found growing near *H. seselifolia* (see plate 10).

H. seselifolia. This dainty, much-branched annual has masses of pure white flowers massed near the tips of the smooth stalks. They sometimes fade to pink. From 7 to 11 stems branch out at ground level and grow to a height of about 22 cm. The tiny, soft hair-like leaves are concentrated near the base of the stems, being divided into 7 to 11 segments that are scarcely 1 cm long. This blooms profusely in spring in the Kamiesberg, on Spektakel Pass and in other places. There are 3 varieties of this species (see plate 10).

There are several other white-flowered species of *Heliophila*, such as *H. amplexicaulis*, *H. laciniata* and *H. variabilis*, which are common in Namaqualand, but the differences between them are confusing and chiefly of interest to botanists.

Springbok Bossie, Springbok Bush

Hertia pallens (Othonna pallens)
(Compositae)

This small woody shrub grows up to 1 metre in height and has long thin branches covered with narrow leaves, about 3 cm long and 3 mm wide. The semi-erect yellow daisies become hairy as they mature. This is a common species around Springbok in Namaqualand and also occurs in the drier parts of the Cape and the southern O.F.S.

Ink-plant, Inkblom, Rooipop, Katnaels, Skilpadblom

Hyobanche sanguinea
Scrophulariaceae)

Bursting through the dry soil like a glowing pomegranate, this attractive flower is a parasite, living on the roots of nearby bushes and plants. It has no real leaves and the flowers form a velvety thick spike of brilliant red flowers up to 10 cm in length. It dies away after shedding its seeds. This plant may be seen in the field around Springbok and in many parts of Namaqualand, as well as in the Karoo and Cape coastal districts (see plate 26).

Other parasites that occur in Namaqualand include the shrubby Mistletoe, *Viscum capense*, with green stems and white berries, that grows on trees and bushes; *Thesium strictum*, a root-parasite with green stems, small narrow leaves and minute starry white flowers, that is generally found on rushes and sedges (Restiones), and *Loranthus oleifolius*.

Lessertia

Lessertia capensis
(Leguminosae)

A sprawling annual, this has tiny, reddish or purple pea-shaped flowers, followed by decorative, slightly inflated seed-pods, which are beige and veined with maroon. The stalks and pinnate leaves are hairy. This little plant may be seen at the foot of Spektakel Pass, west of Springbok.

Leyssera, Vaaltee (Bush Tea)

Leyssera tenella
(Compositae)

A dwarf, dainty annual, this has extremely thin stems clothed with short hair-like leaves. Its tiny branchlets are topped with very tiny thin-petalled yellow daisies, about 2 cm wide. This is found in very sandy, dry soil near Droedap.

L. gnaphaloides has denser leaves and larger flowers. It also occurs in Namaqualand.

Limeum

Limeum africanum
(Phytolaccaceae)

A small perennial with branching stems covered with small oval leaves, this has rounded heads of numerous tiny cream flowers with 3 to 5 petals. These are enclosed by 3 bracts that are conspicuously striped down the centre and add to the decorative quality of the flower-head. This plant is found near Steinkopf and Vioolsdrift, as well as in the S. W. Cape.

Limonium peregrinum (L. roseum)
(Plumbaginaceae)

Sea Lavender, Strandroos

This evergreen shrubby perennial has large heads, fully 12 cm across, of flat, papery pink flowers, that are most attractive when they are in bloom from spring to mid-summer. The plant grows to about 60 cm and has lax branching stems covered with oval, leathery leaves. This species grows wild in the coastal districts of the S. W. Cape and extends to the southern part of Namaqualand.

L. namaquanum is a small woody shrublet with branches that grow to 30 cm. The flower-head is very small and a miniature version of *L. peregrinum*. The small leaves are long, thin and rounded at the tips, about 3,5 cm long and 3 – 4 mm wide. This species occurs only in Namaqualand on stony slopes near Wallekraal.

Manulea
(Scrophulariaceae)

Manulea

There are numerous species of *Manulea* in S. Africa and several attractive kinds occur in Namaqualand. These are annuals with thick taproots that store moisture and enable the delicate flowers to withstand drought. The tiny 5-petalled flowers are grouped in a mass near the top of the stems and the leaves are grouped mainly near the ground in a tuft or rosette.

M. androsacea. This has a tuft of spoon-shaped leaves at ground level and numerous slender stems grow up from the base to about 18 cm, topped by small heads of white flowers. It grows on Spektakel Pass in the shade of rocks.

M. benthamiana is an annual with a thick spike of white flowers with yellow centres. It is scabious-like in appearance. The stalks grow to about 35 cm and the leaves form a basal rosette. This grows in sandy places near Springbok and Port Nolloth, as well as in the S. W. Cape, where it is more common than in Namaqualand. It also occurs in other provinces and S. W. Africa.

M. nervosa is one of the most attractive species, with its delicate mauve and yellow flowers arranged in simple spikes about 7 cm in length. There is also a pale blue and yellow colour form. The plant grows to about 20 cm in height. It may be seen in dry red sand near Pofadder and Aggeneys as well as near Springbok and Steinkopf. This is the most common species in Namaqualand and it extends to Kenhardt in the northern Cape (see plate 24).

Microloma
(Asclepiadaceae)

Red Wax Creeper, Coral Creeper, Kannetjies

These are dainty twining perennials that creep into low bushes for support, bearing clusters of tiny red urn-shaped flowers in the spring, being quite conspicuous because of their colouring. There are 12 species that are exclusively South African and found mainly at the Cape. The following occur in Namaqualand and have red flowers.

M. namaquense may be recognised by its smooth branches and thin simple leaves, almost needle-like, which are scattered along the stems.

M. sagittatum is very common in the fields around Springbok and may be distinguished by the slight hairiness along the branches and by the shape of the leaves, which are partly narrow and partly arrow-shaped or spear-shaped. The flowers are a soft rose-red (see plate 43).

Monopsis *Monopsis campanulata*

(Campanulaceae)

This dainty annual has a flat rounded flower, about 1½ cm across, which is dark violet and shaded pink at the centre. It consists of 5 rounded petals and is almost phlox-like in appearance. The soft sprawling stems have tiny soft leaves growing to one side of the stem and are topped by the flowers. This attractive little plant seems to prefer moist, sandy soil near streamsides and blooms from spring into summer. This flower is not seen in masses, but is most appealing and delicate. It occurs near Springbok and south-wards into the Clanwilliam and Piketberg areas of the Cape (see plate 20).

Wilde Rabas *Monsonia umbellata*

(Geraniaceae)

This plant sprawls on the dry sandy plain east of Springbok, on the road to Pofadder, bearing masses of delicate white 5-petalled flowers, about 2 cm across. The tiny heart-shaped leaves are about 1 cm across. Other species in this family have larger and more showy blooms, especially those in the S. W. Cape, but this has a fragile charm, all the more noticeable because of its arid surroundings. Rabas is a Hottentot name referring to *Pelargonium*, and has been extended to allied plants (see plate 24).

Nemesia, Leeubekkie *Nemesia*

(Scrophulariaceae)

Attractive, slender annuals, there are several species in Namaqualand and numerous kinds in the S. W. Cape and other parts of Africa. Very few are well-known, but all have charm. The flowers have a short, broad tube opening into 2 large lips, with the upper divided into 4 rounded or oblong lobes. There is a single spur or pouch at the back of the flower. The flowers are grouped in branched clusters at the top of the stems and the simple, soft leaves are either entire or serrated along the edges.

N. anisocarpa. A bushy annual growing to 30 cm, this is covered with numerous small dark purple flowers with a yellow centre and a long spur behind each flower. This may be found in hot dry places along the road from Springbok to Pofadder.

N. brevicalcarata may be pure white or streaked with deep mauve. The tiny flowers are half the size of those of *N. versicolor*. A miniature form grows wild in the Kamiesberg and a larger form occurs in the S. W. Cape.

N. macroceras. Delicate salmon-yellow flowers at the tips of slender stems, growing to 35 cm, characterize this annual. It grows in sandy places, shaded by small bushes, on Spektakel Pass and other parts of Namaqualand. It has similar flowers to those of *N. parviflora*, but those of *N. macroceras* are larger (see plate 13).

N. versicolor. The fascinating flowers on this slender annual are generally half yellow and half white, but may also be plain pink or blue, as well as combined with white. They vary a great deal in size and appearance, often growing to 45 cm in height. They flower in spring and early summer and are usually found in sandy places. This species is common in Namaqualand and the S. W. Cape (see plate 13).

Osteospermum Early Morning Daisy
(Compositae)

O. hyoseroides. One of the few daisies that open early in the morning, this also closes early, rolling back its petals early in the afternoon and so reducing its brilliant effect. When fully open, however, it is one of the most spectacular of the annuals that bloom in the spring. It is common on the hills around Springbok, seeming to spring from the stones themselves, but also covers fields with dazzling colour from Kamieskroon to Steinkopf and the Richtersveld.
This tall annual grows to about 60 cm in height and has brilliant golden flowers in clusters at the tops of the branching stems. The flowers measure 5 to 6 cm across and have dark centres. The distinctive botanical feature of this species is the white appearance of the broad bracts behind the flowers, with thin black lines down the centres. This is the most important difference between this species and *O. amplectens* (see plates 2, 4, 36).

O. amplectens could be confused with *O. hyoseroides*, especially as it is a variable annual with dark-centred, deep yellow flowers. The flowers, however, are smaller than those of *O. hyoseroides* and the distinguishing characteristics are the narrow involucral bracts, which are not white and distinctively lined. The leaves vary considerably and may be broader in some cases. This is very widespread throughout Namaqualand and the S. W. Cape.

O. oppositifolium. Large perennial bushes of over 1 metre in height, these have clusters of golden flowers with dark centres, measuring about 4 cm across, at the top of slender stems. The narrow, simple leaves are arranged singly opposite one another up the stems. They are about 4 cm long and 7 cm wide near the base, becoming very thin and short, barely 1 cm long, near the top of the stems. This species may be seen near the top of Spektakel Pass.

O. sinuatum. A low shrubby perennial, this has branched, reddish, woody stems, growing to 60 cm in favourable conditions. The bright yellow flowers have golden centres and measure about 5 cm across. The rough-textured, oval, toothed leaves, up to 5 cm long, are spaced at intervals up the stems. This species may be seen around Springbok.

Othonna Othonna
(Compositae)

Some of the *Othonna* species have very succulent leaves and are collected as succulents by growers, but they have been grouped under perennials here.

O. incisa is a perennial with extremely large fleshy leaves that may be 35 cm long and 12 cm wide. They are sometimes simple and sometimes dentate. The stems are long and slender, standing well above the foliage and branching near the top, with purple daisy flowers, each about 4 cm across, at the tips. This occurs near Springbok and Garies.

O. abrotanifolia is a soft shrub growing to 1 metre, that has broad heads of small yellow daisy flowers, each 2 cm across. The leaves are very finely divided. This species is found in rocky hills around Steinkopf.

Pelargonium, Crane's Bill

Pelargonium
(Geraniaceae)

This huge genus comprises 230 S. African species, many of which are parents of modern hybrids. Pelargoniums are distinguished from true Geraniums by their unequally arranged flowers, divided into 2 upper and 3 lower petals. Those that occur in Namaqualand belong to the section that have succulent, thorny stems or tubers that store water in order to resist drought, and generally lose their leaves during the hot dry summer. They come into growth again in autumn and flower in springtime.

P. echinatum. This has a short, prickly stem, from which several long flower-stalks arise near the ground to a height of about 20 cm. They are topped by clusters of small flowers, each head about 7 cm across. The flowers may be white or pink, marked with crimson spots, or brilliant purple, spotted with dark purple. The broad greyish-green, thin-stalked leaves are roughly triangular, with serrated and scalloped margins, about 3 cm wide and as long. They are numerous near the base of the plant. This species grows on dry, stony slopes from the Kamiesberg to Port Nolloth and the Richtersveld, as well as on hillsides in the drier parts of the S. W. Cape (see plate 12).

P. fulgidum is a shrubby evergreen perennial with large, light green, deeply indented leaves. The clusters of small scarlet flowers are borne above the foliage on dark succulent stems. This species occurs from about 40 miles west of Springbok to the coast at Port Nolloth and Kleinsee.

P. incrassatum. NAMAQUALAND BEAUTY.

Quite the most outstanding *Pelargonium* in Namaqualand, this has thick clusters of brilliant cerise-purple flowers that grow beside the brightest annuals and can be spotted from afar. They generally lie scattered in little clearings among orange daisies in the fields around Springbok and Kamieskroon, where they are plentiful. The plant forms a tuft of large leaves, which are deeply cut and indented, at ground level, and the sturdy flower-stems emerge from these to a height of 20 to 25 cm. Each flower-head is fully 8 cm wide. This species has a small beetroot-like tuber (see plate 12).

P. pulchellum. Similar in appearance to *P. incrassatum*, this also has large, indented leaves near the base, with long-stemmed clusters of flowers that are white with red

PLATE 25

1 *Ornithogalum polyphyllum,* a tall green and white Chinkerinchee.

2 Spiral-leaved *Albuca spiralis* has green and white flowers with characteristic spreading outer lobes.

3 Beetle Daisy, *Gorteria diffusa* subsp. *diffusa,* has extraordinary beetle-like petals. It is spectacular in masses.

48

2

↓ 3

1

2

3

4

PLATE 26

1 Sheltering under their little leaf-umbrellas, the flowers of the Karoo-pampoen, *Radyera urens*, are protected from scorching heat.

2 Desert Spray, *Cadaba aphylla*, is a sprawling shrub of the arid region, which blooms in spring and summer.

3 The glowing root-parasite, *Hyobanche sanguinea*, bursts through the ground in spring.

4 The miniature wild Asparagus, *A. juniperiodes*, resembles a tiny juniper tree. It is found on stony soil in the mountains.

PLATE 27 →

The two conspicuous long "horns" on *Diascia thunbergiana* have given this attractive annual the common name of "Horingtjies", meaning "little horns". It prefers damp soil near streams, but may also be found on dry sand. It grows here together with *Ursinia anthemoides* subsp. *versicolor.*

1

2

1

2

spots. The basal stem is woody and there are corn-like nodules below it. This species occurs on rocky, sandy hillsides near Garies, Kamieskroon, and Steinkopf, as well as southwards to Van Rhynsdorp in the S. W. Cape.

P. quinatum. This twiggy, rough bush grows from 30 cm to 1 metre and is covered with very tiny leaves. It has large single blooms which are cream, veined with purple. This occurs near Garies, as well as at Steinkopf and in the Richtersveld.

Pentzia pilulifera (Matricaria pilulifera) Pentzia
(Compositae)

Many-flowered heads of small blooms characterise this genus and *P. pilulifera* has several small flat corymbs of small yellow disc flowers branching from the top of an erect stem, about 20 cm in height. Each tiny flower is about 1 cm wide, but they are showy in the mass. The delicate, ferny, aromatic leaves are tiny and reminiscent of those of *Ursinia.* This perennial herb is common around Springbok.

P. suffruticosa has even smaller flowers, about half the width, in flattened heads about 8 cm across. It is a less bushy plant than *P. pilulifera.*

Pharnaceum lanatum Droedas Kruiden
(Aizoaceae) (Healing Herb)

This small perennial grows to a height of about 20 cm, with a tuft of very short, hair-like leaves at the base and numerous slender stems forking into hair-thin stems which are topped with a lacy veil of very tiny reddish flowers, scarcely half a centimetre in length. The whole effect is extremely dainty and delicate.

This little drought-resistant plant grows in dry sandy places between Springbok and Pofadder, north of Garies and in dry parts of the S. W. Cape.

P. lineare is a similar species from the S. W. Cape, with larger flowers and longer, thicker leaves, that was used in old Dutch medicines as a healing herb.

Polycarena selaginoides Polycarena
(Scrophulariaceae)

A tiny annual, about 8 cm in height, this branches at ground level and bears tiny, pale blue, 5-petalled flowers near the top of the plant, borne singly at the top of the forked hair-like stems. Each flower is scarcely 1 cm wide and hairy in the mouth. The very short narrow leaves are broader at the tips. This little plant grows on Spektakel Pass

PLATE 28

1 Tiny flowers swaying on tall, slender stems draw one's attention to *Ixia scillaris,* that grows on the rocky hillsides of Spektakel Pass.

2 The dainty *Cyanella orchidiformis* flourishes nearby, but is common from Nama-qualand to the S. W. Cape.

| Karoopampoen, Karoo Pumpkin | *Radyera urens (Allenia urens, Hibiscus urens)* |

Karoopampoen, Karoo Pumpkin — *Radyera urens (Allenia urens, Hibiscus urens)*
(Malvaceae)

A fascinating perennial, this has branches that sprawl on the ground to the extent of about 30 cm, with large rounded leaves up to 12 cm across, standing up on long stalks like parasols, shading the small flowers that lie beneath them from the blistering Karoo sun. Each leaf is thickly felted with white and the stems are equally furry. The tips of the claret-red, bell-like flowers peep from a hairy green calyx. This plant is common in the Karoo, Bushmanland and S. W. Africa, as well as Namaqualand, where it may be seen near Steinkopf and Klipfontein. *Radyera* is named after the distinguished South African botanist, Dr. R. A. Dyer (see plate 26).

Small Pepperbush — *Relhania pumila*
(Compositae)

A small prostrate annual that branches from the base, this has slender stems covered with thin, needle-shaped, sticky-hairy leaves. Yellow-centred yellow daisies, about 2½ cm across, are scattered along the stems. It is common in Namaqualand and S.W.A., extending southwards to Clanwilliam and Ceres.

Wild Scabious — *Scabiosa columbaria*
(Dipsacaceae)

A perennial that grows to a height of 1 metre, this has long-stalked flowers that stand well above the mass of foilage. The flowers may be white, pale blue or mauve. They form small heads, about 2½ cm across, and bloom from spring into summer. The stalked, deeply cut leaves spring from the ground and measure about 8 cm in length. Both the leaves and stems are slightly hairy.

This plant is found in hilly places in Namaqualand and near the coast at Hondeklip Bay, as well as in other provinces and Tropical Africa.

Sebaea — *Sebaea pentandra*
(Gentianaceae)

A bright annual that may be dwarfed or grow to 50 cm, this has branching stems bearing compact heads of yellow flowers at the tips. Each small flower has a tube that opens into 5 oval lobes. The leaves are simple, oval and arranged in pairs opposite one another up the stems. The species occurs in Namaqualand, but there are numerous others which are distributed mainly in the coastal areas of South Africa.

Several species of this vast genus, common throughout the world, help to create the springtime spectacle in Namaqualand. Some of them may be weedy by gardener's standards, but grow in such profusion that they form rich masses of brilliant colour stretching to the horizon during a good season. They usually bear clusters of small daisies that are grouped together in a broad compact head that is showy in the mass. The colours are mainly yellow, but there are several mauve or pinkish flowered species.

S. abruptus. A tall bushy annual, this grows from 45 to 60 cm in height and has individual yellow daisies, about 2 cm across, grouped in heads about 8 cm across. They are brilliant in the mass, covering the foliage when in bloom in spring. The leaves are red-veined and variable, some simple in shape and some deeply indented. This species is common in the fields south of Springbok. *S. inaequidens* is similar, but has smaller heads of flowers (see plates 2, 9, 15, 36).

S. arenarius. This neat annual has yellow-centred mauve or pinkish-purple flowers, each about 3 cm across, grouped in large heads at the top of the bushy plant, which branches at ground level. The leaves are bright green and deeply dentate, being smaller near the top of the plant. The leaves and stems are sticky-hairy and this is a distinguishing characteristic of the species. *S. arenarius* is common in Namaqualand, often seen in association with orange Namaqualand Daisies, as well as in the S. W. Cape and S. W. Africa.

S. cakelifolius. Very similar to *S. arenarius*, and found in the same districts, this has slightly larger pale mauve or deep purple flowers with yellow centres. It can be recognised because the whole plant is smooth in texture. It grows to about 25 cm in height.

S. elegans. This tall bushy annual grows up to 60 cm in good conditions, but varies considerably in size. The mauve or purple form is common in the S. W. Cape, but a pale creamy-sulphur form comes from Namaqualand near Kamaggas. The flowers have yellow centres and the leaves are soft and curly.

S. laxus. Very similar to *S. abruptus*, this has larger bright yellow flowers in spring. The sticky leaves are more deeply divided and more robust in appearance. This species may be seen in stony soil on Spektakel Pass, near Steinkopf and in the Richtersveld, as well as in other parts of the Cape.

S. cinarescens. A shrubby plant that grows to 1 metre, this has large yellow flowers, 3 – 4 cm across. The leaves are massed near the base of the plant and felted with grey beneath. This bush grows near Garies and Kamieskroon and extends to S. W. Africa, as well as the S. W. Cape.

Stipagrostis brevifolia (Aristida brevifolia) Large Bushman
(Gramineae) Grass, Twagras, Bossiesgras

This perennial grass grows up to 1 metre in height and has a much-branched, woody rootstock. The spikelets are yellow, often tinged with purple. This grows in tufts on the arid plains of Bushmanland, east of Springbok, where it may only be knee-high.

Sutera, *Sutera*
Wild Phlox (Scrophulariaceae)

Some of the Suteras are annuals, which will be described here, but 2 others may be found in the section on shrubs in this book.

S. dielsiana. An erect annual growing to 50 cm in height, this has sticky aromatic, serrated, slender leaves, that grow up to 5 cm in length. The large flowers, about 2½ cm long, may be white, yellow, pale blue, mauve or purple, with dark lines leading down into the throat. This grows wild over a broad area in Namaqualand and has been collected from Garies up to Okiep.

S. tristis. AANDBLOM, EVENING FLOWER.

An erect annual that grows to about 30 cm tall, this has clusters of pale yellow flowers near the top of the stem, which are scented at night. Each flower has a long thin tube opening into 5 spreading narrow segments, about 1½ cm across at the top. It occurs in sandy places in the Richtersveld, around Springbok, Spektakelberg and southwards to the S. W. Cape.

Ursinia *Ursinia*
(Compositae)

Among the most brilliant of all the orange daisies in Namaqualand, Ursinias are so rich in hue that they sometimes seem to glow with a reddish cast when they are seen in their millions, covering fields and mountains to the horizon with a blanket of colour. They are particularly showy in fallow fields around Kamieskroon, but also form an outstanding feature of the spring floral display around Springbok and Nababeep. The plants are tallest when there have been good rains, but they are spectacular even when dwarfed by extreme drought.

The different species vary in small botanical details, but all may be recognised by their distinctive light green feathery foliage. The flowers are borne singly on slender stalks that stand well up above the foliage and the drooping buds straighten up as they come into flower. The stiff green scales surrounding the base of the flower-head are characteristic and the seeds are white and papery (see plate 2).

U. anthemoides subspecies *versicolor* (*U. pulchra*, *U. versicolor*). The original species, *U. anthemoides*, occurs in the S. W. Cape, but the subspecies *versicolor* is an annual that is common in Namaqualand. There are many variations in size and colour of flower and plant, but they all have brilliant orange flowers with a dark spot at the base of the petals, forming a narrow mahogany band around an orange centre. This band often becomes broader in cultivation. The flowers vary in size from 3 to 5 cm in diameter. They are often brownish on the underside. The plants themselves vary from a compact form of 20 cm to 35 cm in height (see plates 27, 33).

U. cakilefolia. The bright orange flowers of this annual have no markings in nature, but often hybridise in cultivation and produce a dark band. This species has a dark, blackish-purple centre and the flowers measure about 5 cm across. The scales behind the flower-head are slender, pointed and outlined with purple. The plant grows to a height of 20 to 35 cm.

U. calendulaeflora. A rare, dwarf species, this has an orange flower with black spots near the base of the petals and the bracts are plain green. It occurs in dry places at Bitterfontein and in the Richtersveld.

U. chrysanthemoides var. *chrysanthemoides.* A much-branched shrubby perennial, this grows to a height of 5-100 cm and may be erect or sprawling. The dense, green, ferny foliage has leaflets divided into short, narrow lobes and the stems are reddish. The stalked flower-heads stand well above the foliage and measure up to 4 cm in diameter, with the dark, blackish, central disc about 1 cm across. The narrow ray florets are yellow or whitish above and may be brownish-purple or coppery beneath, or sometimes yellow, but this is rare.

This plant flowers from July to December, but mainly in spring. It is common in sandy soil in the S. W. Cape, and extends from the Eastern Province to Namaqualand, where it occurs from Garies to Okiep. The variety *geyeri*, which has bright scarlet flowers, occurs only in the Ceres district.

U. speciosa. Very like *U. cakilefolia*, this annual has plain yellow flowers with a purplish-black centre, measuring a little over 5 cm across. The scales behind the flower-head, however, are rounded rather than pointed. This is a shorter plant, growing to about 20 cm in height, which is found in dry areas in Namaqualand, S. W. Cape and S. W. Africa.

Wahlenbergia Bell-flower
(Campanulaceae)

These delicate annuals, which are usually pale blue, although some are azure and some whitish, seem almost too dainty to survive the hot dry places in which they bloom. They retain their fresh simplicity in scorching sunshine on baking sand, being borne in masses on slender wiry stems. A few are perennials. The flowers have 5 broad pointed petals, forming a deep cup, and face upwards from the tops of the stalks. The small leaves are alternate and more numerous near the base of the plant. All the species are attractive.

W. acaulis. A miniature plant, this has tiny blue flowers with a white throat and the narrow hairy leaves make a flat rosette on the soil, about 7 cm across. This little annual flowers in sandy stony places from Garies to the Kamiesberg and near Steinkopf.

W. androsacea (*W. arenaria*). This annual has a rosette of leaves lying flat on the ground, from which slender branching stalks arise, bearing very tiny single flowers, scarcely 1 cm in length. Each is a pale blue or white with mauve markings. This species grows wild near Kamaggas and is wide-spread in all provinces of S. Africa and S. W. Africa.

W. annularis. This bushy annual has large pale blue flowers, each $3\frac{1}{2}$ to 4 cm across, and about the same length, at the tops of numerous thin stems that may grow to a height of 35 cm. This is common on the red sands on the road between Springbok and Pofadder, near Aggeneys, flowering in springtime in combination with pale yellow *Grielum* and other small annuals (see plates 15, 23).

W. capensis. A slender annual growing to about 36 cm, the stems branch from the base and bear deep blue flowers singly at the tips. The narrow, indented leaves, about 4 cm long, are grouped near the base of the plant. This species is common in the S. W. Cape and in Namaqualand.

W. oxyphylla. A small, rounded prickly perennial about 30 cm in height, this has minute pale blue or white flowers. It occurs in dry places near Springbok, Okiep and in the Kamiesberg.

W. prostrata. The unusual china-blue flowers, 2 cm wide, on this prostrate plant face upwards to reveal a white starry centre with blue dots. Some are deeper blue and others a delicate mauve. The slender stalks radiate from the centre at ground level. This delicate little annual spreads on the hot red sand in association with *W. annularis* (see plate 23).

Drumsticks *Zaluzianskya*
(Scrophulariaceae)

These small flowers may be recognised quite easily when they are closed, for the tiny petals curl inwards like a ball and the thin flower-tube resembles the handle of the Drumstick of the common name. The petals are generally backed with dark red which becomes apparent when the flower is closed. When open, the petals may be white, yellow or lilac, with a yellow or orange "eye" at the centre. There are 5 heart-shaped petals, deeply notched at the top and joined at the narrow base into a thin tube. The verbena-like flowers are arranged in an open spike at the tips of the stems. The leaves are simple and the plants are usually sticky and hairy.

Z. villosa. An annual that varies from 10 to 30 cm in height, this has compact heads of small white or pinkish-lilac flowers, each with a yellow or orange "eye". They open early in the afternoon and are very sweetly scented. The narrow, tapering and hairy leaves are arranged alternatively up the stems. This plant occurs in dry red sand near Springbok and the Richtersveld in Namaqualand, as well as in the S.W. Cape and S. W. Africa.

Z. gilioides. This small annual grows to about 12 cm and has tiny, thin-stalked yellow flowers which are reddish-brown beneath. It occurs in red sand east of Springbok and south of Kamieskroon.

THE BULBS

PLATE 29

1 The spicy scent of this little Purple Kalkoentjie, *Gladiolus scullyi,* formerly called *G. venustus,* adds to its allure. This is common in Namaqualand, flowering in springtime.

2 This striking large Gladiolus, *G. equitans,* resembles the Kalkoentjie from the S. W. Cape, *G. alatus,* but it is larger and has much broader leaves.

1

2

2

3

Little corms may roll into a rocky cleft and, sprout to flower in the sunshine. The spectacular *Lapeirousia silenoides,* commonly called Springbok Painted Petals, has an occasional white sport.

PLATE 31

1 All Babianas have attractive flowers, even though some are dwarf bulbs. *Babiana dregei* may be recognised by its sharp-tipped leaves and purple or mauve flowers.

2 A natural bouquet of *Babiana dregei* and *Lapeirousia silenoides* emerges from a rock fissure.

3 The rich, cerise flowers of *Babiana geniculata* make this one of the most striking of the Namaqualand species. It was formerly called *B. velutina* var. *nana.*

1

2

3

The Bulbs

The major flower spectacle in Namaqualand is created by annuals and perennials, chiefly of the daisy family, but there are drifts of beautiful and interesting bulbs to be seen, many of which have a special grace of form or colour that is attractive and exciting.

A mass of nodding Tulps (*Homeria miniata*) with sprays of salmon or lemon flowers crowning the tall swaying stems, and slender leaves that sweep down to the ground, has a most graceful quality as well as a colourful effect. To come upon a stand of deep purple Naeltjies (*Lachenalia violacea*) is to experience a thrill, not only because of their rich colour, but because of the charm of the exquisitely formed miniature bells crowded along the sturdy little upright stems. A sloping bank may shimmer with the silky sheen of yellow Sorrel (*Oxalis copiosa*) or a creamy species (*O. obtusa*), while even the slender, dainty *Ixia scillaris*, with its tiny pink or white flowers, will attract one's attention from a passing car.

Isolated bulbs with striking flowers will also catch the eye, like the bright vermilion *Kentrosiphon*, which stands out in solitary splendour from a hillside, or the brilliant carmine-red *Lapeirousia silenoides*, that is so spectacular in spite of its dwarf size.

The collective term "bulb" has been loosely used for all plants that store food during the dormant period and which have rootstocks that are more correctly termed corms, tubers, underground stems or fibrous roots. The climate of Namaqualand, with its soil baked by the sun during summer, is suited to the spring-flowering bulbs which are so abundant in all parts of the world that have rain in winter and a long, dry summer. The spring-flowering bulbs of Namaqualand, however, must endure even drier conditions than most, for they experience a meagre rainfall during the winter months and depend for extra moisture on the mist that drifts in from the nearby ocean, as well as the nightly dews.

Sometimes they are found in the shade of rocks and benefit from the extra moisture conserved at the base. Large bulbs like *Whiteheadia* and *Haemanthus* are found near large boulders. Other small bulbs roll into crevices between the rocks and emerge to flower above them in the sunshine.

Most soils in Namaqualand are sandy, so that even the little available water drains away rapidly and there is no danger of soft bulbs rotting when they become dormant

PLATE 32

1/2 The silken flowers of all Sorrels unfurl in the sunshine, gleaming from a distance. This yellow-flowered *Oxalis copiosa* has heart-shaped leaflets. The variable *Oxalis obtusa* has many forms, including this pink variant, flushed with yellow, which is known as Form B.

3 The Ugly Duckling or Veldskoenblaar, *Massonia latifolia*, is a stemless sturdy bulb that is common around Springbok.

The charming Lachenalias that are so prevalent, and the attractive Chinkerinchees (*Ornithogalum*), have soft white bulbs that would rot very quickly if they were subjected to moisture when dormant. One sturdy white-flowered *Lachenalia* in particular, as yet un-named, blooms bravely in an area with as little rainfall as 20 mm (1 inch) or less in a season.

On the other hand, some of the bulbous plants are water-loving and may be found flowering in a vlei or pan – a shallow depression in heavier soil where water collects in spring, often standing from 10 to 20 cm deep. Here one may see a mass of starry yellow blooms (*Romulea citrina*) that may seem to float on the surface of the water, waving its reflection into little pools.

The following genera are represented in Namaqualand, but many have very few species that occur in this dry area, for they occur mainly in the moister districts of the south-western Cape.

(Liliaceae)

Although these flowers are not very showy, some are interesting and have a quiet charm. The shape of the flower makes the genus easily recognised. It has 6 segments, with the 3 inner segments arching inwards to form almost a tube, while the 3 outer segments swing outwards. Several flowers are arranged at the top of the stem, with the bottom buds generally opening first. Some are small plants and others quite tall. It is extremely difficult to differentiate between the species, which are little known, but some Namaqualand species are distinctive and bloom in spring.

A. altissima. TALL ALBUCA, MAERMAN.

A tall plant growing to a metre in height, this has a strong thick stalk with a spike of flowers, each 2 cm long. They are white with a broad green stripe down the centre of each segment. The broad green leaves are long and fleshy. The large onion-like bulb is white and succulent. This shallow-rooted plant is found in sandy places near river beds in the Kamaggas area.

A. canadensis. This plant varies from one third to a full metre in height. The flowers are yellow with a green band down the centre of the 3 outer segments. The central tube is greenish-yellow, tipped white.

A. cooperi is similar to *A. canadensis*, but is a little shorter, growing from 20 to 50 cm. It can be recognised by the tuft of fibres at soil level, formed by the splitting of its leaf bases and bulb scales.

A. spiralis. SPIRAL-LEAVED ALBUCA.

Growing from 15 – 20 cm in height, this has a loose spike of 4 to 6 flowers, which are yellow, banded with green and up to 2 cm in length. There are 10 or 12 curly leaves, channelled down the face and up to 20 cm in length. There is a small oval bulb, without fibres at the top, in contrast to *A. cooperi*, which has fibres around the top of the bulb (see plate 25).

Androcymbium Anthers-in-a-boat,
(Liliaceae) Bobbejaanskoen,
Patrysblom

Curious little short-stemmed plants, these nestle close to the ground, and are unusual in the Lily family in that they have a corm instead of a bulb, which is buried deeply in the soil. These are several Cape species, with at least 2 in Namaqualand. They are all characterized by having flowers that are insignificant in themselves, but with conspicuous anthers that are enclosed by boat-shaped bracts enfolding one another in a cabbage-like manner. These may be greenish, pinkish-mauve or white.

A. ciliolatum has broad, pure white bracts at the centre, so that the little plants lie scattered over a field like pieces of paper. They are common at Leliefontein in the Kamiesberg mountains. (see plate 33)

Anthericum

See under *Chlorophytum* or *Trachyandra* for plants formerly named *Anthericum*.

Rat's Tail,	*Antholyza*
Rottestert,	
Hanekam	(Iridaceae)

The tribe *Antholyza* was previously separated into several genera, one of which was *Anaclanthe*. This whole genus, *Anaclanthe*, has now disappeared as it is now incorporated by Dr. P. Goldblatt under *Antholyza*. The well-known *Antholyza ringens* is a species from the S. W. Cape, but one other species is now recorded from Namaqualand.

A. plicata is similar in appearance to the striking *A. ringens*, but is a very large plant, varying in height from half to about 1 metre. It has bright red tubular flowers, arranged in a double-sided cluster at the top of the stem, like *A. ringens*, but the notable characteristic of *A. plicata* is the velvety texture of the bracts which hold the flowers. The ribbed leaves are very long and tough, emerging in a tuft at soil level. This species grows in the coastal sand dunes called the Sandveld, along the entire west coast of Namaqualand. Specimens have been found at Port Nolloth and Alexander Bay, as well as in the Richtersveld, where the corms are eaten by tribes who live there.

A. namaquensis is a species that was formerly separated, but is now regarded as a form of *A. plicata*. It is taller, with smaller flowers and a more hairy stem.

| Wild Asparagus | *Asparagus juniperoides* |
| | (Liliaceae) |

Several attractive indigenous species of this large genus are well-known ornamentals both in the garden and in the vase, but these grow wild in the eastern districts of S. Africa. *Asparagus* "ferns", unlike true ferns, are remarkably drought-resistant, especially as some have water-storing rootstocks and occur in dry, hot places in nature. Several kinds occur in Namaqualand, of which the following is probably the most distinctive.

A. juniperoides, which grows in sandy and stony soil on Spektakel Pass in Namaqualand, is a quaint species that is well-named, for it resembles a minature juniper tree. Although grouped in colonies, each little plant emerges singly from the soil, growing to a height of about 21 cm. The soft, needle-shaped foliage overlaps to form a thick plume, 2 to 3 cm in diameter, that tapers at the base and to the top. The tiny white flowers are streaked with green and followed by red berries (see plate 26).

| Babiana, | *Babiana* |
| Babiaantjie | (Iridaceae) |

About 17 species in this large genus occur in Namaqualand, while the remainder of the 80 species are found mainly at the Cape. Their strong colours usually make them conspicuous among other flowers, even if they are dwarf or solitary. All Babianas may be recognised by their slightly hairy, ribbed leaves that appear to be "pleated" and arranged in asymmetrical fans. The flowers may be cup-shaped or irregular and two-lipped, often with attractive painted markings on the lower segments. Many have long tubes and this seems to be a characteristic of the Namaqualand species. The tough fibrous corms are buried deeply beneath the surface, but are often dug up and relished by baboons, from which their name is derived.

Probably the most eye-catching species in Namaqualand is *B. geniculata*, which was formerly known as *Babiana velutina* var. *nana*. This is a dwarf of only 8 to 10 cm, but the rich, crimson-magenta flowers are so showy that they may be discerned easily from the roadside, when travelling by car south of Springbok and near Kamieskroon. This species has long-tubed flowers, with 2 parallel white marks on each of the 2 lower segments. The ribbed leaves are twice as long as the flower stem and arranged to one side of it (see plate 31).

One of the tallest species of this area is *B. striata*, which has a stem of 8 to 12 cm in height, bearing a spike of 3 to 10 pale mauve flowers, with the lower lobes pale yellowish-green or faintly marked with purple. The leaves are up to 12 cm long. This flowers early in June or July, and is found on hilly slopes. Var. *planifolia* differs slightly from the species.

B. fimbriata has an erect, branching stem, with 5 to 8 mauve flowers that have a yellowish tube and are arranged on a one-sided spike. The leaves may be recognised by their spiral twist and are 6 to 10 cm long. This species grows in sandy soil at low altitudes and blooms during August and September.

B. lobata is very similar to the above species, but has more erect and less twisted leaves. The erect stem is up to 12 cm in length and has a loose spike of 7 mauvish flowers.

B. tritonioides is a very tall species, from 10 to 15 cm in height, that is similar to a *Tritonia* in appearance. It bears a one-sided spike of 7 – 10 dull mauve flowers, of which the 3 lower segments are cream, shading to green towards the centre. There are 7 – 8 leaves. This species occurs on sandy plains in the north-western area and near Port Nolloth.

Among the many dwarf species that flower close to the ground, one may recognise the following:

B. attenuata. This has a short stem bearing a one-sided spike of 4 to 9 fragrant flowers. They are blue, with large, pale yellow and white marks above the base. This species is usually found in stony ground at the foot of mountains near Kamieskroon, blooming in August and September.

B. dregei has a short, branched stem bearing a one-sided dense spike of deep purple or mauve long-tubed flowers, blooming in August and September. The 5 – 7 pungent leaves, from 12 – 30 cm long, are rigidly erect, with sharp tips, and sometimes up to 3 cm across. This species is common around Kamieskroon, Leliefontein and other places, growing in rocky crevices at low altitudes (see plate 31).

B. curviscapa is a variable species with a short stem and a one-sided spike of richly coloured flowers, which may be magenta, wine, violet or dark blue, marked with white near the centre. The leaves vary from being short and curved to growing above the flowers. This species grows in sandy soil on flats and low hills around Springbok.

B. falcata has a subterranean stem and blooms at ground level with 2 to 6 flowers emerging from below the red soil on the sandy flats in the north-east of Namaqualand, extending into the O.F.S. and S.W.A. They are fragrant, but khaki-coloured, suffused with mauve and green near the centre. The 6 – 7 leaves are curved and wide, standing erect up to 7 cm.

B. flabellifolia is dwarf species with 5 – 6 purple flowers that are marked with white near the centre and bloom in June and July.

B. framesii var. *kamiesbergensis* is a purple-flowered dwarf species, blotched white on the lower segments, that is found at low altitudes on mountains to the north.

Bulbine, Katstert (Cat's-tail)

Bulbine
(Liliaceae)

Golden yellow flowers forming a dense spike at the top of a sturdy stem characterize this group of plants. There are several species in Namaqualand, varying in size and shape, but all have fluffy stamens on the small open flowers and may be easily recognised by this character. This is the only way in which the layman can distinguish between the flowers of *Bulbine* and *Bulbinella*, which have smooth filaments. The leaves are short and fleshy, being able to store water in dry areas. Several occur in sandy places in the vicinity of Springbok, but botanists are still uncertain about naming the species (see plate 35).

Bulbinella, Katstert, (Cat's-tail), Seeroogkatstert, Swartturk

Bulbinella floribunda (B. setosa)
(Liliaceae)

Rich daffodil-yellow flowers form an attractive "poker" at the top of a sturdy stem. Each glistening flower is shaped like a tiny star with 6 segments and the filaments are smooth, as distinct from the hairy stamens of *Bulbine*. The stem grows to a height of about 60 cm and the long leaves are broad and deeply grooved down the centre. This plant likes moist places and may be seen growing in damp soil on rocky hillsides near Springbok and Kamieskroon or in marshy places where water collects, such as at Grootvlei. There is also an orange-flowered colour form of this species, as well as smaller forms that were formerly named *B. setosa*.

B. cauda-felis (B. caudata) is a similar white-flowered species. These species all bloom in August and September in Namaqualand and also grow wild in the S. W. Cape (see plate 34).

Pennants, Suurknol

Chasmanthe fucata
(Iridaceae)

This rare species grows wild in the Kamiesberg mountains in Namaqualand, emerging from rock crevices in small tufts. It blooms during spring and becomes dormant in summer. There is a flat tuft of stiff leaves and the reddish flowers stand above them in a graceful spike. They are tubular, with protruding stamens. The rootstock is a large flattened corm.

This species is similar to *Curtonus paniculatus* from the eastern Transvaal summer-rainfall area. Both were formerly classed under *Antholyza*.

Hen-and-chickens

Chlorophytum
(Liliaceae)

This huge family is distributed throughout Africa and in many warm countries of the world. The South African species are common from Natal to Namaqualand.

C. undulatum. This common Cape species is extremely variable, according to the circumstances in which it grows. The botanist, Acocks, believes all the forms to be variations of this species.

The plants have decorative tufts or rosettes of tapering leaves, which may be folded, twisted or crisped and grow at varying heights to 50 cm. They emerge from small, tough, drought-resistant rhizomes with tuberous roots. The small white flowers are densely grouped at the top of a slender stem and bloom in late spring.

C. crassinerve is a smaller species with short leaves that are mottled with purple. The small white flowers are evanescent. This species occurs only in Namaqualand, from Garies to Okiep.

Crinum variabile Crinum
(Liliaceae)

An unusual bulbous plant for dry areas, *Crinum variabile* was rediscovered as recently as 1961 by Mr. D. S. Hardy of the Botanical Research Institute, growing in large numbers in the bed of the Groen River near Garies. It is a robust plant with a tall stem, growing to 30 cm, that bears a head of about 9 large short-tubed, lily-like flowers at the top during autumn. These are palest rose-pink, becoming a deep rose as they age. They open wider as they mature, revealing white anthers. The lettuce-green leaves are stiff and narrow, arching upwards and growing to 35 cm in length and about 4 cm in width. This plant is dry in summer, but benefits from the deep waters that gather in the river bed during the winter months.

Cyanella orchidiformis Cyanella
Amaryllidaceae)

To see a patch of *Cyanella orchidiformis* growing amongst the stones on dry soil in the mountains on Spektakel Pass, is to marvel at the fact that they can thrive in such a difficult situation. Each little plant has a dainty, slender stem, growing to 30 cm, with tiny, pinkish-mauve flowers spaced at intervals along its length. The leaves grow in a tuft at the base. This species flowers in September and is found wild from Namaqualand southwards to Clanwilliam in the Cape (see plate 28).

C. hyacinthoides (*C. capensis*) is another species from Namaqualand and the S. W. Cape, which is larger. Each small mauve flower, with bright orange anthers, is held out at the tips of long stalks branching from the main stem, blooming from spring to mid-summer.

Dipidax triquetra Star-of-the-Marsh,
(Liliaceae) Hanekam, Vleiblommetjie

A favourite little plant in the S. W. Cape, where it is common in marshy places and grows in deep water, this is rare in Namaqualand, where it may be seen near Grootvlei, in a damp place at the side of the road, growing together with *Bulbinella*.

The Namaqualand form of *Dipidax* seems altogether more dwarf, probably as a result of dry conditions. It does not produce a tall, many-flowered spike like those that grow in the S. W. Cape, but has a shorter, fewer-flowered spray, growing to about 15 cm in height. The starry white flowers, flushed with pink at the centre, measure about 2½ cm across. The 2 or 3 slender, rolled leaves grow to about 30 cm in length and stand erect behind the flower-spike, spreading outwards slightly. The small bulb lies deeply buried beneath the soil (see plate 34).

Pineapple Flower *Eucomis nana*

(Liliaceae)

This is the only species of *Eucomis* that is found in Namaqualand, and it was formerly known as *E. pillansii*. It grows to a maximum height of 25 cm, but is frequently shorter. It has the typical top-knot of leaves in a tuft at the top of the flower-spike, resembling the top of a pineapple. The flower-spike itself consists of numerous small lily-like flowers, each with 6 segments, which are clustered around the thick central stem. Each flower is pale green and the round ovary in the centre becomes green and decorative as it ages. A rosette of long, green wavy-edged leaves emerges at ground level from the neck of the large bulb, and the flower-stem grows up from its centre, blooming in spring. This plant is green during winter, becoming dormant in summer. There are many other species of *Eucomis* in the summer-rainfall area of S. Africa.

Spider Flower, Spinnekop Blom, Krulletjie *Ferraria*

(Iridaceae)

Although the Spider Flowers are not brightly coloured, they are curiously interesting. Each measures about 4 cm across, with 6 pointed, curly-edged segments. The flowers face upwards at the top of a sturdy branched stem, bearing thick puffed buds, from which new flowers open each day. Long thin leaves arch sideways from the base. There are several species in Namaqualand and in the S. W. Cape.

F. longa has flowers which are pale greenish-grey. It grows in clefts in the rocks in the hills around Springbok (see plate 35).

F. framesii is an attractive, yellow-flowered species that grows among rocks in the northern part of Namaqualand, near Steinkopf.

Blue Sequins, Blousysie *Geissorrhiza secunda*

(Iridaceae)

These little plants are common at the Cape, from the Peninsula up to Namaqualand, blooming in spring or early summer. The flowers are very tiny, scarcely more than 1 cm across, with 6 starry segments, and there are several flowers at the top of thin,

PLATE 33

1 The cup-shaped flowers of *Homeria miniata* are fragrant and the plant grows from 30 – 60 cm in height. It is seen here with *Oxalis copiosa, Arctotheca calendula* and *Ursinia anthemoides*.

2 Like little pieces of white paper scattered over the hard soil, these dwarf bulbs are common near Leliefontein. This is *Androcymbium ciliolatum* and it is commonly called Anthers-in-a-Boat.

2

PLATE 34

1 Dwarf *Romulea citrina* thrives best in moist sandy soil and is common from Kamieskroon to Springbok.

2 *Bulbinella floribunda* (*B. setosa*) produces its pokers in moist despresions, generally on hillsides.

3 Star-of-the-Marsh, *Dipidax triquetra*, is rare in Namaqualand, where its growth is modified by drought.

PLATE 35 →

1 *Trachyandra patens* may be recognised by its prim, brown-striped flowers and curly leaves.

2 One of the succulent species of *Bulbine* (Katstert), often seen in Namaqualand and identified by the fluffy stamens.

3 This Spider Flower, *Ferraria longa,* grows among rocks near Springbok.

1

2 3

1

2

wiry stems that grow to 30 cm. Although small, they are a conspicuous deep bluish-purple. The leaves are wiry and thin, but they vary in size according to the locality in which they are found.

G. *namaquensis* is another species with larger, longer-tubed, deep blue flowers and the stem grows to about 26 cm in height, branching near the top. The narrow leaves are strongly ribbed and pointed. This species starts blooming early and the flowers persist into late spring. It grows among stones in dry places from Klipfontein to the Richtersveld.

Gethyllis Koekemakranka
(Amaryllidaceae)

These curious plants are uncertainly named because of the fact that the leaves appear above ground during the winter months and the evanescent flowers appear during summer after the foliage has become dormant. The aromatic, many-seeded, finger-like fruit pokes up through the soil in the autumn, when the first rains begin, and is edible, being sought by children as well as animals. Unless one sees all three stages of growth, it is difficult to name the plants correctly. Several species occur in the Cape with two in Namaqualand. They all bear the old Hottentot name, "Kukumakranka," which is now spelt "Koekemakranka".

The best-known species is G. *spiralis*, which occurs throughout the Cape Province, western O.F.S. and S. W. Transvaal.

G. *britteniana*, a Namaqualand species from the Richtersveld, is described in Flowering Plants, plate 1428. It has a large bulb, about the same size as that of a daffodil, which is covered with papery white scales. There is a rosette of narrow spiralled leaves, which are fleshy in texture, about 17 cm long and 3 mm wide. The flower appears just above the ground and has six creamy white segments, which are broad and tapering, and six bunches of yellow stamens which stand erect at the centre.

Gladiolus Gladiolus, Kalkoentjie
(Iridaceae)

The large and important *Gladiolus* genus is well represented in S. Africa, but only a few occur in Namaqualand. They all belong to the "Kalkoentjie" group of wild *Gladioli*, which name means "little turkey", with reference to the drooping lower segments. All have hooded flowers.

G. *arcuatus*. AANDBLOM.

This species has dull mauvish-purple flowers that are very fragrant in the evening. The plant has a short, wiry stalk and very thin leaves. It is found on sandy soil among boulders in many parts of Namaqualand.

PLATE 36

1 A broad flowery landscape with a hillock of golden *Osteospermum hyoseroides* and fields of yellow *Senecio abruptus.* Tiny cerise *Lapeirousias* and other bulbous plants speckle the pathway.

2 The miniature *Moarea fugax* has an Iris-like flower with beautiful sculptured outlines and a single wiry leaf.

G. equitans (formerly *G. alatus* var. *namaquensis*).

This is best described as a giant form of the Kalkoentjie (*G. alatus*) from the S. W. Cape. It is the most striking of the Namaqualand species, having 9 or 10 flowers at the top of a 30 cm stem and short, curved leaves which are very broad, up to 4 cm wide. The flowers are bright orange, with 3 broad, spreading upper segments and 3 drooping lower segments, which are yellowish-green at the base. It blooms in September and is found on rocky hillsides from the Kamieskroon area to the Orange River.

G. orchidifloris. GREEN KALKOENTJIE.

Dull greenish-yellow and white hooded flowers make this a species of interest to an enthusiast. There are about 6 smallish flowers at the top of a 50 cm stem. It occurs from Kamieskroon to Steinkopf, among bushes on sandy soil.

G. salteri. A rarely seen species, this has small greenish flowers.

G. scullyi (*G. venustus*). PERSKALKOENTJIE, PURPLE KALKOENTJIE.

This delightful little *Gladiolus* is attractive because of its strong spicy scent. Although the flowers are small, about 4 to 5 cm in length, they have dainty pointed segments and attractive colouring, being yellow and tipped with bright purple. There are about 3 flowers loosely arranged on a 30 cm stalk and the leaves are thin and grass-like. This species blooms in September and is very common in the S. W. Cape as well as in Namaqualand, blooming in thousands near Nababeep and Spektakel Pass in a good season. (see plate 29)

Torch-lily, Paint-brush | *Haemanthus namaquensis*
(Amaryllidaceae)

An unusual plant in a dry area, *H. namaquensis* was discovered among the hills of northern Namaqualand, finding shade and moisture between the boulders. The vivid flower-head is composed of a mass of pink filaments that are crowded together to form a brush and surrounded by dark red bracts. This is at the tip of a sturdy stalk growing to 8 cm in length. The 2 long strap-shaped leaves are about 9 cm wide and grow up to 40 cm long. The leaves grow during the winter months, after the plant has flowered in autumn.

Evening Flower, Aandblom | *Hesperantha*
(Iridaceae)

Delicate flowers that open in the afternoon only, often towards sunset, characterize this genus. They have 6 segments and are often fragrant. These plants are dormant in summer.

H. angusta. This species is also known as *H. bachmanii*, which is a synonym. Although the flowers are white, they grow near the top of a 30 cm stem and make an attractive display when seen in the mass. The segments are wide-open and reflexed, while the leaves are slender. This species grows in the mountains near Wuppertal and is common in Namaqualand.

H. flexuosa (*H. namaquensis*) is a slender plant with small white or pink pendulous flowers. The tube, 1 cm in length, opens into 6 small segments, with the stamens just showing at the opening. This species has 2 or 3 thin leaves, up to 10 cm in length, and a branching stem, growing to about the same length. It grows in stony soil in the Kamiesberg and near Springbok.

H. pauciflora has mauvish-pink flowers that are red on the outer surface. It is one of the showiest in the group, bearing 2 to 4 flowers near the top of the branching stem in the spring. Each segment is about 2½ cm long and almost 1 cm wide. The stem grows to 25 cm and there are several firm, sword-shaped leaves up to 16 cm in length.

Hexaglottis flexuosa   Hexaglottis
(Iridaceae)

This genus is not very well-known. The flowers of *H. flexuosa* are golden-yellow, sometimes dotted at the base of the narrow segments, and the outer lobe may be reddish or lined with green down the centre. They are characterized by a strong, unpleasant scent and are fairly numerous in flattened spikes at the top of a branching stem, that grows from 20 to 60 cm high. They open only in the afternoon. There are 1 to 3 narrow long leaves on the plant.

This species grows in hard, dry soil in the sunshine on the lower slopes of the mountains near Garies, Kamieskroon and Springbok. It also grows wild over a large area of the S. W. Cape. (See Lewis in Journal of S.A. Botany 25:233.1959)

Homeria miniata   Homeria, Rooitulp (Red Tulip)
(Iridaceae)

This species is common in Namaqualand, growing in sandy soil in dry areas, as well as in moist soil near streams. It usually has pale salmon-pink flowers, but may also be cream coloured, and has a pale yellow "star" at the centre. The flower forms a cup-shape, measuring about 3 cm across, and is fragrant. This species differs from *H. breyniana* (*H. collina*), which is confied to the S. W. Cape, in the shape of its flowers, for the latter are larger and the petals widespread. *H. miniata* has clusters of flowers at the top of the branches and reaches from 30 to 60 cm in height, blooming in August and September. The long, green whip-like stems arch over to trail on the ground (see plate 33).

H. schlechteri is a common, yellow-flowered species that grows to the same height as *H. miniata* and is often seen in the same locality.

H. herrei (*H. spiralis*) has dark violet cup-shaped flowers about 4 cm across, and thin, very spiralled leaves, growing to 30 cm in length.

Hypoxis aquatica   Star Flower
(Amaryllidaceae)

This plant has small white starry flowers, about 2 cm across, with the outer segments pale green below. The leaves are very thin. It is found in the Kamiesberg mountains, near Leliefontein.

Wand-flower, Corn Lily, Kalossie (Little Bell), Ixia

Ixia

(Iridaceae)

Only 3 of the 30 species in this exclusively South African genus occur in Namaqualand. Although these are not amongst the showiest in the genus, they have a delicate grace and may be noticed while travelling during the warm hours of the day, when their small, bowl-shaped flowers are wide-open. They are usually grouped near the top of a slender, wiry stem and, when in bud, resemble an ear of corn. (For scientific descriptions see "The Genus Ixia" by G. J. Lewis).

Ixia latifolia var. *ramulosa*. These have loose groups of about 3 deep purple flowers, which may sometimes be mauve, each about 2 cm in length, blooming in August and early September. They are generally seen half-open and appear to be funnel-shaped from the side. The stem grows to about 30 cm in height. The sturdy bluish-grey leaves are about half the height of the stem. This species may be seen from Garies and around Kamieskroon to Springbok.

I. rapunculoides is a complex, very variable species, with local forms that are so distinct that they have been treated as varieties. Two of these varieties are found in Namaqualand.

Var. *rapunculoides* has a few small, delicate pale blue or mauve flowers on a tall stem that grows from 25 to 55 cm high. The 3 or 4 leaves are much shorter. This variety blooms in August and September and is found at high altitudes on the Kamiesberg mountains.

Var. *namaquana* is a short-tubed form with larger flowers that may be purple or bluish-mauve on a 30 cm stem. It also occurs in the Clanwilliam and Calvinia districts of the Cape, but the typical Namaqualand form has longer-tubed flowers that give it a superficial resemblance to *I. latifolia*, although it has much shorter stamens and short, curved leaves. It blooms in August and September on dry, rocky hill-slopes near Steinkopf and Springbok.

I. scillaris. Although the flowers of this species are small, there are from 7 to 25 on a tall stem that grows from 20 – 50 cm high, so that they are easily noticed. They vary from pale to deep pink or magenta and are occasionally white or mauve, with a greenish stain at the centre, and are faintly fragrant. There are 3 to 7 leaves that vary in length, about a third of the height of the stem. This species may be found in bloom from August to November in the Cape as well as Namaqualand, where it is seen in rocky places on Spektakel Pass, near Springbok (see plate 28).

Red Trident

Kentrosiphon saccatus (K. propinquus)

Iridaceae)

Even a single specimen of this striking plant that grows on a rocky hillside is spectacular. It has a tall branching stem with a series of about 20 bright vermilion tubular flowers at the tops of the branches, blooming in September. Each flower is 2 – 3 cm long and slightly hooded. The leaves are long and narrow. The plants vary in height from 45 – 70 cm according to the locality in which they are found. These were formerly given separate names, but they are now grouped under the single name of *K. saccatus*. The most robust form occurs in Namaqualand and the S. W. Cape, while a more slender form from S. W. Africa is considered a subspecies *steingroeveri* (see plate 38).

Lachenalia (Liliaceae) — Cape Cowslip, Viooltjie, Naeltjie

To come upon a colony of stocky little Lachenalias, blooming bravely on sandy slopes or sprinkled between bright annuals, is to be charmed by their daintiness and pretty colouring. These are probably among the most attractive and characteristic bulbs in the Namaqualand region.

Of the 65 species that are exclusive to South Africa and southern South West Africa, most are found in the Cape, with about 6 in Namaqualand, while about 10 are still to be described. Although some have been collected, their names cannot be used at this stage. One particularly robust white-flowered species grows west of Springbok in dry red soil on the way to Bushmanland.

Lachenalias from the south-west Cape may be golden or cimson, but those in Namaqualand vary from pure white, pale yellow, through delicate opalescent blues and mauves to a deep violet. All are characterized by tubular flowers, arranged in a spike on a fleshy stem that grows to about 20 – 25 cm (8 – 10 inches) in height. The flowers are tubular, some up to 3 cm long, while others are bell-shaped and barely a third of that length. The 3 outer segments form a tube and the 3 inner protrude beyond these, often with different colouring, so that the whole effect is richly variegated. There are generally 1 or 2 short, fleshy leaves, sometimes short and rounded or sometimes rolled and grass-like. The small fleshy bulbs lie close to the surface of the soil and are baked dry during the hot summers.

The named Namaqualand species are as follows:

L. anguinea. A little-known species, this has short, whitish, stalked, pendulous bells, tipped green, about 1 cm in length, with exerted stamens. The flowers are ranged on a purple-spotted stem, 15 – 30 cm in height. There is one very long broad leaf, a little longer than the stem, which is barred with darker green on the back.

L. hirta has a slender stem to 20 cm, with a few-flowered spike of short cream and opalescent blue flowers, and a single narrow leaf, clothed with bristly hairs, clasping the base of the stem.

L. mutabilis, is a dainty plant to 17 cm, with distinctive short iridescent, bluish-mauve flowers, tipped with bright yellow and deepening to crimson with age. The top buds taper to form a delicate spray and the lightly spotted leaf is broad and tapering, with wavy edges. The typical form of *L. mutabilis* has bright blue sterile buds near the top, but there are several forms. One that grows near Nourivier in the Kamiesberg has flowers that are bluish-mauve at the base, with yellow inner segments, and are brown-tipped.

L. ovatifolia can be recognised easily by its crowded, rich mauve flowers, which are tipped with brilliant purple, and 2 fleshy green leaves that are almost rounded when flattened. This robust species is frequently found in moist, sandy soil near Springbok, flowering in August and early September (see plate 13).

L. unifolia. While the typical *L. unifolia* occurs mainly in the S. W. Cape, two or more variations may be seen near Kamieskroon and Klipfontein. Both have thin stems with a lax spike of flowers and may grow from 20 to 35 cm in height. While one has sky-blue flowers, shading into brown and tipped with lemon yellow, with a mottled maroon stalk and green leaves streaked with white at the base, the other variation has pale mauve flowers with lemon tips and is striped with purple where the short, narrow leaf clasps the stem.

L. violacea is a tall plant that grows up to 30 cm and the spike is crowded with numerous, bluish-green, stalked flowers tipped with brilliant purple. The inner segments are deep violet, with exserted stamens. The leaf is narrow and half folded down the centre, spotted with purple at the base. It occurs near Droedap (see plate 13).

Painted Petals *Lapeirousia*
(Iridaceae)

There are some attractive little plants in this genus, which is being revised by Dr. P. Goldblatt. Approximately 9 species occur in Namaqualand, but some of these are not well-known. *L. silenoides* (*L. speciosa*) is certainly the most spectacular and frequently seen.

L. barklyi has 3 – 5 slender-tubed, lilac flowers on a 15 cm stem and a single grass-like leaf.

L. fabricii (*L. anceps*) has 2 – 5 white or cream, slender-tubed flowers at the top of a stem that grows to 30 cm. There is one ribbed leaf, about 15 cm long. This species occurs from Piketberg northwards to Steinkopf and is common in Namaqualand near Okiep.

L. silenoides (*L. speciosa*). SPRINGBOK PAINTED PETALS.

This plant grows only to 10 cm in height, yet the dense clusters of little carmine flowers are so bright and gay that they can be seen from afar. The 3 lower segments have darker red patches and a cream mark at the base. A white sport may be seen occasionally. The flowers grow at the top of flattened stems that are overlapped by waxy-textured bracts. This species blooms in August and September near Garies, Kamieskroon and Springbok, sometimes growing in large patches in sandy soil between bushes and sometimes in rocky fissures (see plates 30, 31, 36).

Ledebouria, Wild Squill *Ledebouria (Scilla)*
(Liliaceae)

There are 16 species of *Ledebouria* in South Africa, formerly classed under *Scilla*. These usually have small flowers appearing at different times from the leaves and they are, therefore, very difficult to identify. The species illustrated is probably *L. undulata*, with its long, ruffled-edged leaves that grow through winter and spring. They are about 10 cm in length and 2 cm wide. The short flower-spike, about 5 cm in height, has very tiny greenish flowers and appears after the foliage has died down, some bulbs flowering in early and some in late summer. The bulb is comparatively large, about the size of a daffodil. *L. undulata* grows in red soil on the coastal belt in Namaqualand (see plate 38).

Massonia latifolia
(Liliaceae)
Veldskoenblaar, Ugly Duckling

More curious than beautiful, this plant is common in the vicinity of Springbok and southwards to Kamieskroon. It also grows as far east as Aliwal North in the northern Cape. It is stemless, with two broad leathery leaves growing flat on the ground, each about 10 cm long, with a small flower-head nestling like a tuft between them. This is very short and seldom more than 4 cm in diameter, consisting of greenish-white flowers that are composed mainly of stamens. They bloom in the spring and the plants are dormant in summer. *Massonia* grows in sandy soil in hot place (ssee plate 32).

Melasphaerula ramosa
(Iridaceae)
Baardmannetjie

A straggling plant that grows to about 50 cm in length, this has thin branching stems bearing a cloud of dainty white flowers in springtime. They have 6 pointed petals and are usually streaked with mauve at the centre. The narrow leaves emerge in a rosette at ground level and become dormant during summer. This is a common plant at the Cape and it extends into Namaqualand.

Moraea
(Iridaceae)
Morea, Tulp

M. fugax (*M. edulis*). This lovely little miniature plant grows to a height of about 12 cm and the white iris-like flower appears in spring at the top of a thin stem. There is one long, wiry leaf arching from the base. The three lower segments of the flower are finely pencilled with grey and marked with yellow and the small erect crests are crisply sculptured. This little species grows amongst annuals and bushes on the lower slopes of the hills around Springbok. This is possibly the most attractive of the few species that occur in Namaqualand (see plate 36).

M. arenaria (*M. framesii*) is a very common species with fleeting yellow or whitish flowers. They are in clusters of 2 to 4 at the top of a stem that grows to 12 cm. The leaves are very thin, tiny and curly.

M. bolusii (*M. spiralis*) also has fleeting yellow flowers, about 3 cm across, and grows to about 12 cm in height. It has a long single leaf, about 1 cm wide, that is curiously crisped along the whole length. This species is found near Okiep.

M. gawleri (*M. crispa*) has small, fleeting lilac or yellow flowers, about 2 cm in length, arranged in a long spray. There are 2 long leaves, each 15 – 30 cm long and less than 1 cm wide, which are generally crisped along the edges. The tiny corm has a thick tunic of fibres. This species occurs in the S. W. Cape as well as in Namaqualand.

M. serpentina has pale yellow, mauve or white flowers. The short, thick leaves are very spiralled, which is a drought-resisting feature.

Chinkerinchee *Ornithogalum*

(Lilaceae)

This large family is well represented in South Africa, as well as in Europe and Asia. Few are well-known, except the popular cut-flower from the S. W. Cape, O. *thyrsoides*. Most species have white, starry flowers, often striped with green, while there are a few spectacular, rarer kinds with yellow or orange flowers. Some species have a dense tapering spike of flowers, while others have a flattened shape. The 6 segments of each flower are wide-spread, opening slowly and generally remaining in bloom for several weeks.

A few species grow wild in Namaqualand, being found in sandy soil, especially on stream banks. This dries out in summer so that the soft bulbs growing near the surface can bake dry in the hot sunshine. Most of these species have insignificant white flowers, striped with green.

O. *pruinosum* is one of the most showy and common species. It is a tall plant, growing to 40 cm, with a dense spike of white flowers that have a small green mark at the base of the segments. The broad, bluish-green leaves grow in a short tuft at soil level. It may be seen in valleys below Spektakel Pass.

O. *xanthochlorum* is a large, coarse plant with large broad leaves that grow in a rosette near the base. The tall stem grows to 60 cm, bearing a dense tapering spike of flowers that are more green than white. It occurs north of Springbok.

O. *polyphyllum*. This has a tall stem, 30 – 60 cm in height, bearing fragrant flowers for about half its length. These are whitish or yellow with a green stripe down the centre of each segment, which is up to 15 mm long and 5 mm broad. There are numerous long thin rolled leaves, which are flattened near the base and measure up to 25 cm in length. The large bulb has rough, leathery outer tunics. This plant grows in the vicinity of Springbok (see plate 25).

O. *suaveolens* is similar to O. *polyphyllum*, with white flowers that are striped with green, but it has only 2 or 3 short, broad leaves, which are a distinguishing feature.

Sorrel, Suring *Oxalis*

(Oxalidaceae)

This huge group of plants is common at the Cape, where fields and roadsides may be covered with its delicate, silken flowers of many kinds, chiefly in yellows, pinks or white. There are several species growing wild in Namaqualand, as mentioned below, where the tiny bulbs lie dormant during hot, dry summers and start growth during the winter months. The little plants have typical sorrel leaves that are generally arranged in threes, with heart-shaped or oblong leaflets. The flowers form a shallow cup of 5 overlapping petals that open fully only in bright sunshine.

PLATE 37

1 Looking northwards from a rocky hillside near Springbok, one surveys a flower-strewn valley with the Copper Mountains or Koperberg on the horizon. Simon van der Stel discovered copper here in 1865.

2 A striking large bulb that seeks moisture in the shadow of a rock, *Whiteheadia bifolia* is called Elephants-Ears by local children.

2

1

2 3

1

←PLATE 38

1 Fiery scarlet flowers of the tall Red Trident, *Kentrosiphon propinquus*, are conspicuous against the grey rocks.

2/3 This miniature bulb has its leaves and flowers at different seasons and is, therefore, difficult to identify. This is probably *Ledebouria undulata*, formerly *Scilla*.

PLATE 39

1/2 Azure flowers are rare in nature, so that this pretty shrub attracts attention in the spring, especially as it grows in dry, rocky places in the mountains. This species of Agtdag-Geneesbos, *Lobostemon pearsonii*, may be seen on Spektakel Pass.

2

1

2

O. copiosa. Bright yellow flowers, produced abundantly, characterize this plant that grows wild near Kamaggas and Spektakel Pass, west of Springbok. It has heart-shaped leaflets (see plate 32, 33).

O. flava is a robust succulent plant with 2 to 12 narrow leaflets spreading like a fan from the top of the leaf-stem. The corolla, about 3 cm long, is spreading and may be bright yellow, pale lilac or pink. It is common in Namaqualand and the S. W. Cape, flowering in winter.

O. namaquana. An uncommon species, seen on the banks of rocky streams in the Kamaggas area, this has bright yellow, large buttercup flowers up to 3 or 4 cm across, standing above the compact foliage that forms a bushy plant about 20 cm high. The tri-foliate leaves have oblong leaflets.

O. obtusa. There are so many variations in this species that it has been called a "group-species", for its flowers may be pink, brick-red or pale yellow and the leaflets vary from narrow to heart-shapes, sometimes even on the same plant. It can always be recognised by the fact that all the stems bear backward-pointing hairs and the styles and stamens are well exserted from the corolla tube. They are not treated as varieties, but called Form A. One variant, with pale pink flowers flushed with sulphur-yellow, has been called Form B. *Oxalis obtusa* is widely distributed from Namaqualand to the southern and eastern Cape and the Little Karoo, flowering from June to October (see plate 32).

O. purpurea is another wide-spread and variable species which has transparent dots or streaks on the hairy, heart-shaped leaflets and other small characters. (F.P.A. pl. 1323). A pale salmon-coloured form is common in Namaqualand and there are pink, mauve, rose-purple and white forms elsewhere in the Cape Province.

Romulea citrina Romulea, Frutang, (Iridaceae) Satin-flower

This dwarf plant has several short wiry stems, about 10 cm in length, each bearing a large, cup-shaped flower with 6 starry segments that spread to about 3 cm in diameter. They are bright yellow and face upwards, forming brilliant clusters of flowers nestling close to the ground in springtime. The thin grass-like leaves arch above the flowers. This plant is often found in moist sandy soil in low-lying depressions or in the shallow water at the side of streams, which dry out during summer when the plants are dormant. *R. citrina* is common around Springbok and Grootvlei, near Kamieskroon (see plate 34).

R. atranda var. *luteoflora* has a similar bright yellow flower, but the throat is marked with reddish-brown. It occurs on mountain-slopes in the Kamiesberg near Leliefontein and may be seen southwards in the S. W. Cape as well as in the Karoo, where pinkish-mauve forms occur.

PLATE 40

1 A river of blue *Felicia tenella* flows between orange *Arctotis fastuosa,* rippled with tiny bulbs, and laps against the foot of a bush-covered hillside.

2 Wild Broom or Fluitjies, meaning "little flutes", is a shrub that is covered with pea-shaped flowers in springtime. This species, *Lebeckia sericea,* is probably the most commonly seen in Namaqualand.

Stars, Sterretjies	*Spiloxene (Hypoxis)*

Spiloxene (Hypoxis)

(Amaryllidaceae)

Two species of this genus that occur in Namaqualand, *S. serrata* and *S. scullyi*, have tiny yellow flowers, scarcely broader than 1½ cm and not to be compared in beauty to the large *S. capensis* of the S. W. Cape. Seen in the mass in the wild, however, they have a dainty appeal, blooming in springtime. They open fully in the midday sun, closing in early afternoon.

S. serrata has yellow flowers, backed with green, and extends from the Karoo into Namaqualand.

S. scullyi has slightly larger yellow flowers and occurs in sandy, rocky soil west of Springbok.

Trachyandra, Hottentotskool, Hottentot's Cabbage — *Trachyandra (Anthericum)*

(Liliaceae)

This genus of about 30 species was formerly classed under *Anthericum*, but the Namaqualand species of *Anthericum* are now reclassified under *Trachyandra* or *Chlorophytum*.

Trachyandra may be recognised by the layman in that the starry flowers are always striped down the centre of the 6 segments. A few notable species that are found near Springbok include the following.

T. ciliata is a tall plant growing to 50 cm, with white flowers arranged in a long spike. They are striped with green down the centre. The short leaves are arranged in a tuft at ground level.

T. falcata (*Anthericum drepanophyllum*) is probably the most striking, standing out among the fields of daisies around Springbok and Komaggas. It is called Hottentot's Cabbage or Hottentotskool, referring to the fact that the fleshy, broad green leaves, often about 10 cm wide, were used by tribesmen to cook with their meat. The tall stem grows to about 60 cm and branches into spikes that are crowded with white flowers, striped with brown (see plate 1).

T. patens is a short plant to 15 cm that has interesting leaves, which are thin, rolled and extremely curly, growing in a tuft at ground level. The delicate white flowers have a thin brown stripe down the centre of each segment (see plate 35).

Veltheimia — *Veltheimia capensis*

(Liliaceae)

The bladder-like green fruits of this plant, which are mottled red, are almost as striking as the pale pink or spotted pink flowers .These are tubular and form a compact head at the top of a purplish, sturdy stem, growing to about 30 cm in height. The stem arises from the centre of a rosette of curly-edged, bluish-green leaves. This plant grows in the shade of rocks in the hills around Springbok and in the Kamiesberg. All the species from the S. W. Cape, previously described under different specific names, are now grouped under *V. capensis*, as distinct from the eastern Cape species, *V. bracteata* (*V. viridifolia*). (See Flowering Plants of Africa, plate 1356).

Whiteheadia bifolia   Elephants-Ears
(Liliaceae)

An unusual and striking plant, this is reminiscent of *Eucomis*, but lacks the characteristic "top-knot" of that genus. A sturdy stem emerges from between two large, rounded, succulent pale green leaves that lie flat on the ground, suggesting the common name. There is a thick spike of green flowers, each forming a green cup of white stamens nestling in a pointed, spreading green bract. The flower-spike grows to a height of 10 to 35 cm, becoming taller as the seeds mature. It blooms in the spring, flowering in August and September. It grows in the hills around Springbok near large rocks that provide shade and moisture (see plate 36).

Zantedeschia aethiopica   White Arum Lily,
(Araceae)   Pig Lily, Varkblom

This well-known species occurs in moist soil throughout the S. W. Cape, and extends into all provinces of S. Africa. It is seldom found in dry areas and is rare in Namaqualand, but may be seen in rocky kloofs where small streams trickle down hillsides. It is generally much smaller than those plants growing in marshy soil near Paarl, Stellenbosch and Cape Town, but will grow to two-thirds of a metre in height if conditions are favourable.

The pure white spathe has a bright yellow spadix in the centre and the clump of arrow-shaped leaves is bright green. It is amost too well-known to need description. The plant dies down during the hot summer months.

THE TREES
AND SHRUBS

The Trees and Shrubs

Trees are so scarce in Namaqualand that townspeople who go out for picnics into the countryside have learned to look for the shade of a large boulder rather than that of a tree.

There are trees, however, which are generally to be found in dry, sandy river-beds. Drought-resistant Acacias are most common, but there is also the Dawip (*Tamarix usneoides*) that has tiny, cypress-like leaves that are adapted to survive drought. There is an indigenous Wild Fig (*Ficus ilicina*) that grows very large, but this is so distinctive and unusual that everyone seems to have noticed the few existing specimens.

Large trees that are seen near farm houses generally prove to be the introduced Pepper Trees (*Schinus molle*) and Gum Trees (*Eucalyptus*) that grow easily in dry areas. The Wild Tobacco (*Nicotiana glauca*) is a small tree or large shrub from South America that has naturalised itself all over Namaqualand, as well as in other parts of the country.

The best area for trees is in the moister Kamiesberg range, where lovely small trees like *Freylinia* may be found, but these are uncommon and certainly not typical of the country as a whole.

Succulents which have tall, tree-like shapes include the Kokerboom (*Aloe dichotoma*) and the rarer *Aloe pillansii*, while the larger Euphorbias are more like shrubs than trees.

People who prefer shrubs with attractive flowers, foliage or distinctive habit may find few types to interest them in Namaqualand, but several are outstandingly decorative and will attract the attention of the most discriminating traveller. Some are so attractive that they have already been grown by gardeners, such as *Erica plukeneti*, *Nymania capensis*, *Sutherlandia frutescens* and several kinds of *Melianthus*. Others, like *Lebeckia sericea*, *Didelta carnosa* and *Hermannia stricta* are worthy of cultivation.

On the whole, however, the majority of small shrubs and shrublets are much less than one metre in height and more interesting botanically than horticulturally. Most of them have tiny, narrow leaves that enable them to combat loss of moisture through evaporation. They are most plentiful on hill-sides, where there is a greater precipitation of moisture than on the plains and where they benefit from partial shade from projecting rocks and slopes that cast shadows.

The reason for including so many shrubs and trees in this book is to highlight the surprising number of different kinds that exist in this seemingly desert area. The amateur botanist will be intrigued by the amazing variety of species and may be able to find the detailed descriptions in the following pages useful in helping to distinguish one shrubby plant from another. It is also most interesting to note how many genera of well-known trees and shrubs from other provinces are represented by other species in Namaqualand.

Thorn Tree, Doringboom *Acacia* (Leguminosae)

Only two well-known species in this large genus occur in Namaqualand, where they may be seen in dry river-beds or sandy places. Although they survive in the driest of situations, there may be underground water present.

A. giraffae, the Camel-thorn (Kameeldoring) is a semi-deciduous tree with a spreading crown, growing to 20 metres in the best circumstances, but it is considerably shorter in Namaqualand, where it occurs near Springbok. It is also common in dry areas of the northern Cape, O.F.S., the northern Transvaal and S. W. Africa, in the Kalahari. It has pompon-like yellow flowers in spring, followed in autumn by large, broad, curved pods, each about 10 cm long and almost half as wide. The pointed thorns are in pairs, up to 5 cm long and often swollen at the centre. The short feathery leaves grow to 5 cm in length and are divided into pairs of small leaflets.

A. karroo, the Sweet-thorn or Soetdoring, is the most common species in South Africa and occurs in all provinces, as well as in S. W. Africa and Zambia. It has heavily-scented yellow pompon flowers which are massed near the ends of the branchlets during summer. It is variable in size, but has a spreading shape and grows to about 5 metres in Namaqualand. It has very long paired thorns which are often bleached white during winter when the trees are bare. The leaves grow to about 10 cm in length and are divided into paired leaflets. The dark brown seedpods are long, narrow and often curved, constricted between each seed (see plate 45).

Agathelpis *Agathelpis dubia (A. angustifolia)* (Scrophulariaceae)

A slender, woody, branching shrublet, this has stems growing to 45 cm, which are covered with tiny heath-like leaves. The flowers are extremely tiny and are ranged in a spike near the tops of the stems, sometimes as long as 30 cm. The flowers may be a port-wine or yellowish-brown colour, and each one is not more than 1 cm long, but has a very thin tube opening into 5 lobes. This shrublet is common in the mountains of the S. W. Cape and has been collected in the Kamiesberg in Namaqualand.

PLATE 41

1/2 A spectacular shrub with dainty, glowing flowers, the Desert Rose or Hongerplant, *Hermannia stricta*, is all the more conspicuous because it grows in semi-desert regions of searing heat.

3 The Klapperbos, *Nymania capensis*, is a drought-resistant shrub with inflated, ornamental fruits.

2

3

PLATE 42

A small bush with tiny flowers, *Hermannia cuneifolia* is one of the most common of the yellow-flowered species that thrive in very dry places.

Bright crimson flowers and glossy foliage make *Melianthus pectinatus* a handsome shrub. It is commonly called Kruidjie-roer-my-nie, meaning Touch-me-not, as the leaves have a pungent odour when handled.

Dwarfed by drought, the Balloon-Pea, *Sutherlandia frutescens,* sprawls from a crevice at the base of a boulder.

1

2

3

PLATE 43

1/2 The Red Wax Creeper, *Microloma sagittatum*, is a twining perennial that creeps into low bushes for support. The tiny flowers are also called Kannetjies, meaning "little cans". It may be found in the fields near Springbok.

3 Wild Tobacco, *Nicotiana glauca*, is a large weedy shrub from South America that has naturalised itself throughout Namaqualand, as well as in other parts of the country.

1

Anisodontea triloba (Malvastrum grosulariifolium) Wild Mallow, Bergroos
(Malvaceae)

The small, pale mauve, red-streaked flowers on this shrub resemble those of *Hibiscus*. They have 5 broad petals that overlap slightly to form a shallow cup and the whole flower measures about 5 cm in diameter. The rough, broad leaves have 3 rounded lobes and vary from 1 to 5 cm in length. This open-branched bush grows to about 1½ metres in Namaqualand, where it occurs from Garies to Leliefontein in the Kamiesberg. It also occurs in the dry interior of the S. W. Cape and Karoo, where it may reach 3 metres in height (see plate 22).

Asclepias fruticosa Shrubby Milkweed, Wild Cotton, Gansies, Melkbos
(Asclepiadaceae)

A narrow shrub that may grow to about 3 metres, this has pale green inflated pods in late autumn, which often persist into spring. They are oval in shape with a "beak" at the tip and covered with soft prickles, growing up to 7 cm in length. The small white or yellowish flowers are insignificant, and the leaves are smooth, tapering and vary in length from 5 to 15 cm. This is a wide-spread weed in all provinces of S. Africa and S. W. Africa and has been collected near the Orange River in Namaqualand.

Berkheya fruticosa (B. incana) Prickly Sunflower Doring Gousblom
(Compositae)

This is a shrubby perennial growing up to about a metre in height and covered with very prickly, holly-like leaves. It has decorative large yellow daisies, about 4 – 5 cm across, encircled by spines. This drought-resistant plant is common in sandy places in Namaqualand, from the Richtersveld southwards to Hondeklip Bay and the S.W. Cape.

B. spinosissima. An exceedingly prickly bush, this has dentate leaves with numerous long sharp spines along the edges. The deep yellow flowers, 4 – 5 cm across, are encircled by masses of long spines on the bracts. This is common in the drier parts of Namaqualand, where there are several other similar species (see plate 46).

Boscia albitrunca Shepherd's Tree, Witgatboom
(Capparidaceae)

If ever a tree seemed to form a natural Bonsai, it is this small tree in the stony, arid area near Steinkopf in the north of Namaqualand, where it becomes gnarled and attractively bent in its struggle for survival. It occurs near Steinkopf, Stinkfontein, and along the course of the Orange River into the O.F.S., as well as in the Transvaal, Natal and Tropical Africa.

PLATE 44

1 Shepherd's Tree or Witgatboom, *Boscia albitrunca*, is a small tree in Namaqualand, gnarled and dwarfed in the stony region near the Orange River.

2 Bergviool or Mountain-Violet, *Brachycarpaea juncea*, is an exquisite shrublet with scented flowers.

In typical form it has a whitish slender trunk with a spreading rounded crown and grows to 7 metres, but is half that size in the stony semi-desert. The simple leaves grow to 5 cm and are rounded near the top, tapering to the base. The small, pea-shaped fruits become pale yellow and are edible (see plate 44).

Mountain-violet, Bergviool, Blou Riet — *Brachycarpaea juncea (B. varians)*
(Cruciferae)

Sweetly-scented, deep violet, mauve or white flowers make this plant extremely attractive. The flowers have 4 petals, each about 1½ cm long, and are grouped in showy spikes near the top of the slender, delicate stems. These are so crowded, however, that they form a rounded mass of flower-stems spreading about 60 cm in width. The bush grows to a height of about 1 metre, but is generally shorter, and the thin stems branch from near the base of the plant, becoming woody near the ground. Nevertheless, it has the appearance of a perennial rather than that of a shrublet. The soft, short needle-like leaves are scattered along the stems.

   B. juncea grows in sandy ground on rocky hillsides in many parts of the S. W. Cape, as well as in Namaqualand, in the Kamiesberg and on Spektakel Pass. This is the only species in the genus (see plate 44).

Swartstorm, Desert Spray — *Cadaba aphylla (C. juncea)*
(Capparidaceae)

A spreading shrub that grows to almost 1 metre in height, this has smooth Broom-like branches with very tiny lime-green leaves that are grouped in threes. The crimson flowers are cup-shaped and velvety, with a long protruding red style and branching stigma, divided into 8 red segments, each tipped with yellow. They are grouped in clusters along the stems. This is a plant found in sandy places in the Karoo and northern Cape, extending to Steinkopf in Namaqualand (see plate 26).

Wild Gentian, Christmas-Berry — *Chironia baccifera*
(Gentianaceae)

An evergreen shrublet that grows to 45 cm, this forms a much-branched rounded bush covered with narrow, fleshy, light green leaves. Numerous small, starry bright pink flowers, about 2 cm across, appear in spring and are followed quickly by scarlet pea-sized berries, persisting into summer. There is a prostrate, seaside form with shorter thickened leaves.

   This plant grows on sand-dunes in the coastal area of the Cape and Namaqualand, as well as in the Kamiesberg.

Chrysocoma coma-aurea Karoo Bush
(Compositae)

A branching shrublet that grows to a height of about 36 cm, this has masses of large yellow disc flowers, each about 1½ cm across, massed in heads at the tops of the stems. The leaves are tiny and needle-shaped. This drought-resistant plant is found in the Richtersveld and on the Kamiesberg.

C. peduncularis, from the Kamiesberg, is a more slender and delicate plant with disc flowers that are half the size.

C. tenuifolia is known as a poisonous Karoo bush that spoils grazing. It has yellow thistle-like flowers and needle-like leaves on a small bush to 36 cm.

Cliffortia strobilifera Kammiebos
(Rosaceae) Vleibos
Swamp-Bush

This large shrub grows up to 2 or 3 metres and has grass-like leaves. The bush is much-branched and twiggy and the narrow tapering leaves have rough edges. The minute flowers are insignificant, but the plant is curious because of the large cone-like galls on the branches and twigs, which are composed of broad scales. This is a common shrub in swampy places in all parts of S. Africa and it occurs in river-beds in Namaqualand.

Codon royeni Prickly Bush
(Hydrophyllaceae)

This fierce-looking thorny shrub, about 1½ metres in height, is conspicuously covered with short white needle spines all over the stems, leaves and bracts surrounding the flowers. The tubular flowers are blue, or white striped with purple, and form a deep cup-shape, about 3 cm long and half as broad. This species occurs in rocky places near Garies, Okiep and in the Richtersveld, as well as in the drier parts of the S. W. Cape and S. W. Africa.

Colpias mollis Colpias
(Scrophulariaceae)

A woody much-branched shrublet which is fairly rare, this forms a brittle bush bearing large light yellow, scented flowers, measuring about 2½ cm across. Each flower has 5 petals with a short tube and 2 small pouches in front. The soft, triangular leaves are deeply toothed and about 1½ cm wide.

This little shrub is found only in Namaqualand on granite rock-faces facing south near Nuwerus, the Buffels River and in the Richtersveld.

Crotolaria effusa Bird-Flower
(Leguminosae)

A sprawling perennial that spreads on the ground, this has bright yellow pea-flowers. The leaflets have rounded tips and are grouped in threes. This species is found near Springbok and northwards to the Richtersveld, where a very dwarf form occurs.

Didelta, Perdebos (Horse-Bush)
Didelta spinosa
(Compositae)

A woody shrub that grows from 1 to 3 metres in height, this has a rounded shape and is covered with fresh-looking broad, oval, green fleshy leaves. There are thin spines at the tips and edges of the leaves that are not always apparent. The bright yellow daisy flowers, 9 cm across, are scattered all over the bush and nestle in a circle of leaves at the tips of the branchlets like a posy. Although this is a rugged bush, it has fresh bright colouring and is most attractive in the spring, especially as it grows in hot, stony country on hillsides in the Kamiesberg and on Spektakel Pass, where it is plentiful (see plate 46).

D. carnosa is a woody shrub that is much shorter and sometimes very dwarf. It has orange flowers and may be recognised by its long, narrow rough leaves. Both stems and leaves are succulent in texture. This species grows wild near Garies and Port Nolloth.

Elephant's Foot
Dioscorea elephantipes (Testudinaria elephantipes)
(Dioscoreaceae)

Regarded as one of the curiosities of the plant world, this climber has a tuberous root and a woody stem that is much enlarged at the base above the ground, with raised square portions on it, so that it resembles an elephant's foot. The climbing stem arises from the centre and has slender branches that form a tangled thicket, often growing to 1 or 2 metres in height. The leaves are rounded and grow at intervals up the stems. The small flowers are insignificant and are borne in sprays in the axils of the leaves.

The Elephant's Foot has been plundered in nature because of its fascination for plant collectors and because it contains *Diosgenin*, one of the ingredients of natural Cortisone, but it is now one of the plants protected in nature and its removal would involve a heavy fine (see plate 56).

Diosma, Wild Buchu
Diosma hirsuta (D. vulgaris)
(Rutaceae)

An evergreen aromatic shrub that is reminiscent of *Erica*, this grows to about a metre in height and is covered with short, needle-like leaves. The tiny white flowers are clustered near the tips of the stems and have 5 tiny petals and a short tube. This plant is common from the Cape Peninsula to Namaqualand.

Jakhalsbessie
Diospyros ramulosa (Royena ramulosa)
(Ebenaceae)

A large erect shrub that grows to 3 metres in height, this has tiny dark green oval leaves, about 1 cm long, arranged alternately up the stems. The female plants have small white insignificant flowers, followed by tiny berries. This shrub grows on the hills south of Springbok.

Dischisma Dischisma
(Scrophulariaceae)

These little plants have dense spikes of white flowers which may be confused easily with those of *Struthiola*, unless one examines the shape of each tiny bloom. Those of *Dischisma* have thin tubes which are split in front and dilated into 4 small lobes, whereas those of *Struthiola* have 4 distinct pointed segments with a starry appearance, belonging to the family *Thymelaeaceae*.

D. ciliatum. This small shrublet grows to 45 cm high and its numerous stems are covered with narrow, pointed leaves up to 3 cm long. The small white flowers, each about 1 cm long, are crowded into dense spikes, about 15 cm in length, at the tips of the branches in early spring and summer. This species is common around Springbok as well as in the S. W. Cape.

D. clandestinum. An annual plant growing to 30 cm, this has tiny white flowers densely crowded into spikes at the ends of the stems. The greyish-green leaves are simple, narrow and up to 3 cm long, clasping the stem at the base. This species grows near Kamieskroon.

Dodonaea viscosa var. augustifolia (D. thunbergiana) Ysterhout, Sandolyf, Sand-Olive
(Sapindaceae)

This evergreen shrub grows to over 1 metre and has aromatic, narrow, pointed leaves. Clusters of small, winged, papery seedpods persist on the branches for a long period. This variety comes from the S. W. Cape and can be seen in Namaqualand on the road from Garies to the Kamiesberg.

Dyerophytum africanum (Vogelia africana) Wild Plumbago
(Plumbaginaceae)

A shrubby perennial or undershrub that grows to about 60 cm, this has white flowers that resemble Plumbago. The small oval leaves become slender and pointed at each end. It is found on the hills around Springbok.

Elytropappus rhinocerotis Renosterbos (Rhino-Bush)
(Compositae)

This is one of the more insignificant small bushes that cover the hillsides in Namaqualand and are extremely drought-resistant. This species forms a juniper-like bush growing to 60 cm, and has tiny heath-like leaves covering the branches. There are tiny purple fluffy flowers in the spring. It grows wild from Garies to the Kamiesberg.

Erica, Heath *Erica*

(Ericaceae)

Only two Heaths in this huge genus of over 600 species in Southern Africa, *Erica plukeneti* and *E. verecunda*, have ever been recorded from Namaqualand, according to an authority on this genus, E. G. H. Oliver, of the Botanical Research Institute.

E. plukeneti, the Tassel Heath or Hangertjie, has attractive flowers that bloom most freely during winter and spring, from April to September, as well as at other times during the year. They may be recognised easily by the corolla tube, which is swollen or inflated at the base and narrow at the mouth. The flower, almost 2 cm long, hangs downwards and a tassel of long thin brown anthers, 12 mm in length, protrudes from the mouth to droop downwards. The flowers vary in colour from carmine or purplish-red to pink, reddish-orange or white with green tips. They are grouped near the tops of the stems, drooping downwards to form a loose spike. The whole bush grows to 60 cm and the stems are thickly covered with overlapping needle-shaped leaves.

This is one of the best-known and most widely distributed of all Cape Heaths, occurring on hills and plains from Mossel Bay in the east to Namaqualand in the north, where it is found in the mountains around Springbok and in many localities south-wards. Specimens with especially large and showy flowers occur in the Spektakel area and have been reported from Port Nolloth to Kleinsee.

A. verecunda is an erect shrub to 1½ metres, which has small white or pale-pink, urn-shaped flowers, 3 – 6 mm in length, with conspicuous red stalks. These emerge from the same point and hang downwards to form many-flowered clusters. The leaves are short, narrow and overlapping. This species occurs only on the Kamiesberg range, according to Mr. Oliver, who has furnished me with its distinguishing characters.

Wild Rosemary, Kapokbossie *Eriocephalus africanus (E. umbellulatus)*

(Compositae)

A rigid, aromatic shrub, that grows from 1 to 2 metres in height, this has soft, greyish-green leaves that are short, narrow and crowded up the branching stems. The white flowers appear in clusters at the tips of the branches from mid-winter to spring. Each small flat flower measures just over 1 cm across and is quickly followed by white fluffy seed-heads that resemble tufts of cottonwool.

This shrub is wide-spread in rocky places in Namaqualand and occurs in the Cape, mainly near the coast.

Sumoe *Erythrophysa alata*

(Sapindaceae)

A shrub that grows to about 2 or 3 metres in height, this has rigid branches, dividing into many short branchlets, with small red flowers that are arranged in a large flattened head. Each asymmetrical, tubular flower has 4 petals and hairy protruding stamens. These are followed by decorative, inflated, purplish-red pods, growing in threes, that are slightly pointed and measure about 4 cm in length. The short leaves are divided into small oval leaflets, each about 3 – 4 cm long.

The plant, called *Sumoe* by the Namaquas, was first discovered during van der Stel's expedition to Namaqualand and drawn by Claudius. They found it in sandy areas in the mountains west of Springbok and reported that the fruit was bitter and astringent. It also occurs in other mountainous parts of Namaqualand.

Euclea pseudebenus — Cape Ebony, Ebbehout
(Ebenaceae)

One of the rare taller trees to be seen in Namaqualand, this grows to 8 metres in height and is found in dry river-beds in the Richtersveld and near Port Nolloth, as well as in the drier parts of the northern Cape. It has thin pointed leaves, about 4 – 5 cm in length and the male and female parts are on different trees. The flowers are tiny and urn-shaped, but insignificant.

E. lancea is more common and is a shrub that grows to almost 2 metres. The leaves are broader, about 1 cm wide, and up to 5 cm long. It grows wild in the Kamiesberg and drier districts of the S. W. Cape.

Euryops tenuissimus subspecies *tenuissimus* — Resin Bush, Daisy Bush
(Compositae)

Several species of this genus from the Cape are now grown in gardens, notably the Clanwilliam Daisy (*E. athanasiae*).

E. tenuissimus subsp. *tenuissimus* (*E. kamiesbergensis*) is found in Namaqualand near Hondeklip Bay and forms a shrub, about 1 metre in height, covered with long, soft, needle-shaped leaves. It bears broad heads of small yellow flowers, 3 cm across, in springtime. There are 3 other species in Namaqualand, but this is the showiest.

Ficus ilicina (F. guerichiana) — Wild Fig
(Moraceae)

A large tree, all the more striking because of the few large trees seen in Namaqualand, this grows to a height of about 7 or 8 metres and forms a rounded shape. A huge specimen may be seen on the road to Droedap, south of Springbok, and another at the foot of Spektakel Pass. It occurs on rocky hillsides near Steinkopf and in S. W. Africa. The simple fleshy leaves are about 9 cm long and 2½ – 3 cm broad, tapering at both ends.

Freylinia lanceolata — Freylinia
(Scrophulariaceae)

A small evergreen tree with heavy clusters of small, scented flowers at the ends of the branches in spring, this is one of the more attractive trees seen in Namaqualand, but it is found only in the Kamiesberg, where the rainfall is higher than that of the plains. This species also grows near streams in the mountains of the southern and eastern Cape.

The flowers are pale orange and each has a narrow tube opening into 5 lobes, the whole flower being about 1½ cm in length. The long, narrow willow-like leaves are scattered along the soft branches. This tree usually grows to about 5 metres in height where conditions are favourable.

| Gnidia, Kannabast | *Gnidia geminiflora*
(Thymelaeaceae) |

A sprawling untidy shrublet with yellow or creamy flowers, this may grow up to a metre in height. It has small, heath-like leaves and the tubular flowers generally have 4 small lobes. This species is found near Kotzerus and a few others also occur in Namaqualand.

| Desert Rose, Hongerplant | *Hermannia (Mahernia)*
(Sterculiaceae) |

H. stricta. An outstandingly beautiful evergreen shrub, all the more remarkable because it blooms in such hot and arid surroundings, this is covered with deep, rose-red, dainty flowers in springtime. It grows to about a metre in height, its many-branched woody stems forming a rounded bush. They are covered with small leaves, 7-20 mm long and 2-7 mm broad, that are slightly toothed along the edges. The funnel-shaped flowers, about 3 cm in length, have 5 flaring petals and hang downwards like Fuchsias from the ends of the branches, springing singly from the axils of the leaves. The seedpods have 5 pairs of spreading, curved, hairy "horns" (see plate 41).

H. stricta is found in a broad band on both sides of the Orange River, from Kakamas westwards through Namaqualand to Alexander Bay. It is often confused with a Karoo species with similar rose-red flowers, *H. grandiflora*. This, however, can be distinguished by the stalked flowers springing in pairs from the same point at the ends of the branches, by the deeply indented leaves and by the seedpod, which has no horns.

H. cuneifolia. This is one of the most common of the numerous yellow-flowered types of *Hermannia*. It is heavily browsed by sheep, but grows to 60 cm in good conditions. It forms a spiky, woody shrub covered with tiny wedge-shaped leaves that are silvery-hairy on both sides. Silvery scales also cover the calyx of the small mustard-yellow flowers that droop downwards all over the bush. This species is wide-spread in Namaqualand, the Karoo and the Cape (see plate 42).

H. desertorum is similar, forming a wiry bush. It has more slender leaves, smaller yellow flowers and occurs in the same areas.

H. trifurca is a dainty bush to 30 cm, with rosy-mauve or violet flowers, each about 1 cm long, near the top of the stems. The leaves are narrow and hairy. This species is found near Hondeklip Bay.

PLATE 45

1 Although it is smothered with small pretty flowers, *Sutera ramosissima* emits a disagreeable smell. It nestles among rocks on hillsides and can be noticed from afar.

2/3 The dry bed of the Buffels River is tinted pale mauve with great masses of the dwarf perennial *Felicia scabrida*. The bare tree is *Acacia karroo*, which has yellow pompon flowers during summer. A diamond mine may be seen in the background.

2

3

1

2

PLATE 46

1 The rugged Perdebos, *Didelta spinosa*, is resplendent in springtime.

2 Prickly Sunflower, *Berkheya spinosissima*, is exceedingly spiny.

PLATE 47 →

1 Low patches of sparkling Mesembs (*Lampranthus comptonii* var. *angustifolius*) occur near Springbok, forming a perfect colour combination with yellow *Senecio* and bands of orange daisies that spread to the hills.

1

2 3

Hirpicium — Daisy Bush
(Compositae)

H. alienatum. A small woody shrub growing to 30 cm, this is covered with rough, pointed prickly leaves. The yellow daisies, about 3 cm across, have no markings on them.

H. echinus. This is a less woody, more dwarf plant that is almost like a perennial. It has large yellow daisies, about 3-4 cm across, with black spots in the centre of the thin petals and a broad yellow centre. They occur singly at the tip of rough reddish stems. The short, rough, hair-like leaves have tiny white prickles near the base. This drought-resistant plant is reminiscent of *Ursinia* in appearance and grows in sandy soil on the road to Droedap.

Indigofera — Pea-Flower, Wild Indigo
(Leguminosae)

A large and variable group of shrubs and perennials, these are often distinguished by small botanical details seen under a microscope, presenting many problems to taxonomists. The leaves are generally arranged in threes.

I. spinescens is a small rounded bush of about 40 cm, that may be recognised by the presence of short woody spines. The flowers may be pink, pale red or purple and the very tiny leaves are hair-thin and arranged in threes.

Kiggelaria africana — Wild Peach, Spekhout
(Flacourtiaceae)

This evergreen tree is found chiefly in the forests of the Cape, as well as in other parts of South Africa, where it grows to about 10 metres in height. It occurs at high altitudes in the Kamiesberg mountains of Namaqualand.

The tapering leaves resemble those of a peach and the female has an easily recognised fruit. This is a buff-coloured globe which splits open in the autumn to expose the bright orange seeds.

Kissenia capensis (K. spathulata) — Kissenia
(Loasaceae)

This erect, prickly shrublet, growing to about 60 cm, has white flowers with 5 long thin petals, about 4 cm long and 1 cm wide. The large poplar-like leaves have 3 toothed lobes. This species is the only one in S. Africa and is found in the northern Cape near Pella, Helskloof in the Richtersveld and S.W.A.

PLATE 48

1 The Dew-flowers have glistening, glassy dots sprinkled over their leaves like icing sugar. Deep mauve flowers glitter on the cushions of *Drosanthemum hispidum.*

2 The only species with large white flowers, *Lampranthus densipetalus* has a radiant simplicity.

3 The dazzling red flowers of *Drosanthemum speciosum* stand above the foliage on wiry stems. They have a ring of white or beige at the centre.

Wild Broom, *Lebeckia*
Besembos, (Leguminosae)
Fluitjies

There are many species of these shrubs dotted among the fields of annuals in Nama-qualand and speckled on the mountain passes. All have spikes of yellow pea-shaped flowers and broom-like stems, with their leaves divided into 3 leaflets. The stems are sometimes reedy and therefore named Fluitjies, meaning "little flutes".

L. cinerea. A twiggy bush of over 1 metre in height, this has small yellow flowers and light green hairy leaves. It is found on the sandy coast near Port Nolloth, extending northwards to S.W.A. and southwards to Van Rhynsdorp.

L. cytisoides. This rounded shrub that grows to 2 metres has large, bright yellow flowers, fully $3\frac{1}{2}$ cm long, in 15-cm spikes standing well up above the foliage. The narrow hairy leaflets, each about 2 cm long, are variable in size. This species occurs near Springbok and Kamieskroon, as well as in the S. W. Cape.

L. linearifolia. A rounded grey shrub, growing to 2 or 3 metres, this has twiggy branches and very tiny narrow leaves. The small pale yellow flowers are marked with rusty red and they appear in short spikes in springtime. This occurs on the banks of the Orange River in the N. Cape and in Namaqualand, as well as near Springbok.

L. mucronata. This small twiggy bush has very tiny narrow leaves and small yellow flowers. It is found on sandy dunes in the Richtersveld, as well as in the N. Cape and O.F.S.

L. multiflora. The pale green twiggy branches of this shrub, growing from 1 to 2 metres in height, bear thin, almost hair-like leaflets grouped on threes at the top of very thin stems. The tiny flowers are yellow. This shrub grows all along the western portion of Namaqualand and extends into S. W. Africa, becoming more twiggy and densely leafy as it is occurs northwards.

L. sericea. Probably the most common species in Namaqualand, this forms a rounded bush of 1 metre or more. The flowers stand up in erect spikes all over the bush and may be bright or pale yellow or flushed with pink. The long narrow silvery leaflets are grouped in threes. In general, this species has slightly smaller flowers on shorter spikes than that of *L. cytisoides*, is more densely branched and silky-hairy (see plate 40).

L. spinescens. A very twiggy bush growing to 45 cm, this may be recognised by the sharply pointed twigs protruding from the very tiny leaves. Several long-stalked leaves spring from one point on the woody stems, each with 3 tiny leaflets. The tiny flowers, greenish or creamy yellow, are faintly scented and arranged in short spikes. This species occurs in dry places throughout Namaqualand and the N. Cape.

Namaqualand *Leucospermum alpinum*
Pincushion Flower (Proteaceae)

This is the only species of *Leucospermum* that is known to come from Namaqualand, where it is found solely on the higher peaks of the Kamiesberg Range, according to the monograph by Dr. J. P. Rourke.

It is a small, rounded shrub that grows up to 1½ metres in height. The tiny flower-heads are scarcely 2 cm across and almost spherical in shape. The minute flowers are pale pink and fluffy, with styles that become claret-coloured as they mature. It resembles a *Serruria* rather than the usual Pincushion Flower. Greyish-green in colour, the leaves are thick and tough, with an oval shape that is narrower at the base.

Lobostemon (Boraginaceae) — Agt-Dag-Geneesbos Eight-Day-Healing Bush

L. fruticosus is probably the best-known of this group of shrubby plants, as it is seen so frequently in the S. W. Cape, bearing its blue or pink funnel-shaped flowers on bushes at the roadsides in springtime. It is less common in Namaqualand, but may be seen on Spektakel Pass. It was used medicinally by the colonists in early days at the Cape.

The bushes grow to about 1 metre or more in height and the long stems are covered with rough hairy leaves that are broad and rounded at the tips. The leaves clasping the 2 cm blue or pink flowers are extremely hairy.

L. pearsonii is an attractive rounded bush, about 60 cm high, which is covered with clusters of azure-blue, funnel-shaped flowers, each about 1½ cm long, and flushed with pink at the base. The leaves are very distinctive, as they are narrow, pointed and have rasped edges, but are not hairy. This species occurs on Spektakel Pass, in rocky, sandy ground (see plate 39).

Loranthus oleifolius (Loranthaceae) — Loranthus, Vuurhoutjies, (Matches)

A parasite on trees, this has attractive scarlet flowers with green tips. The green leaves are arranged opposite one another on the long stems and measure 5 cm in length and 3 cm in width. The berry-like fruits contain a sticky substance so that the seed is spread by birds wiping it off their beaks on to branches. This species occurs near Kamieskroon.

Lotononis digitata (Leguminosae) — Lotononis

A sprawling perennial with a deep tap-root, this has a spidery appearance because of its hair-thin twigs and leaves, that are grouped in threes. The small yellow pea-shaped flowers are grouped in clusters at the tips of the branches. It is found sprawling over rocks in the hills between Springbok and Kamieskroon, as well as in the Kamiesberg.

Melianthus (Melianthaceae) — Kruidjie-roer-my-nie, Herb-touch-me-not, Honey Flower

The best-known species in this genus, *M. major*, does not occur in Namaqualand, but there are 3 other shrubs that are common in dry river beds and sandy places around Springbok and to the west towards Hondeklip Bay. All have bladder-like fruits with shiny black seeds. The deeply-cut, ferny-looking leaves are generally pungent when handled and give rise to the common name of Touch-me-not. The large deep red flowers are clawed and unequal in shape, usually arranged in a spike.

M. comosus may be distinguished by its hairy, grey-green, deeply indented and ruffled leaves.

M. minor is a rounded shrub growing to just over 1 metre in height, with small greyish-green leaves that are about 10 cm long and divided into serrated leaflets. The terracotta-coloured flowers nestle below the leaves in springtime and are not very showy.

M. pectinatus is the most attractive of the Namaqualand species, for the spikes of crimson flowers stand well up above the foliage all over the rounded bush in spring. This grows to 1½ metres in height, forming a rounded shape. The deep green leaves are divided into narrow glossy leaflets. The decorative seedpods are mottled with green and red (see plate 42).

Blombos *Metalasia muricata*

(Compositae)

A common shrub in the Cape, this grows to a metre or more in height, and is found in dry hills and bushy plains up to Namaqualand. The erect branching stems are covered with short, narrow, heath-like leaves and tipped with tufted heads of flowers which are usually white, but may be an attractive shade of pink. This bush forms part of the "fynbos" or "macchia" on the Cape mountains and is drought-resistant.

Wild Lucerne *Monechma*

(Acanthaceae)

M. mollissimum. A small woody bush covered with simple rounded leaves, this has small purple flowers, each with a long thin tube opening into 2 lips with 5 lobes, 3 being on the lower lip. This species occurs in hot dry places in the northern part of Namaqualand, near Steinkopf and the Richtersveld.

M. divaricatum is an undershrub from Kamieskroon with heath-like leaves and small purple flowers.

Peperbos, Pepper-bush *Montinia caryophyllacea*

(Saxifragaceae)

A slender soft shrub growing to over 1 metre in height, this has tiny white flowers, with male and female on separate bushes. It has small oval fruits. The simple leaves are smooth and tapering and the form growing in Namaqualand and S. W. Africa has much narrower leaves than that growing in the eastern Cape. This is the only species in the genus and it occurs in the Cape coastal districts.

Steekbos (Prickly-bush) *Muraltia rigida*

(Polygalaceae)

A rigid densely branched shrub that grows to about 40 cm, this differs from the well-known Cape species, *M. heisteria*, in that the leaves are not spine-tipped, as well as in small botanical differences in the flowers. It has small fleshy leaves and tiny, pinkish winged flowers all along the branches. This species occurs in the shelter of granite boulders on the lower slopes of the Kamiesberg and near Leliefontein.

Myrica quercifolia
(Myricaceae)

Wax Berry, Wasbes

A small evergreen shrublet that grows up to 1 metre in height, this has male and female flowers on separate bushes. The female has small reddish flowers amongst the densely crowded leaves, forming a spike at the tips of the branches, which are followed by bluish-grey, rough-textured berries that are hard and waxy on the surface. The numerous stems on this species are closely covered with small deeply indented leaves, like those of an oak. This species is found in Namaqualand, but several others occur in the Cape, Natal and the Transvaal. Other species are found in all parts of the world.

Nicotiana glauca
(Solanaceae)

Wild Tobacco

Although this is an introduced weed from S. America, it has naturalized itself so freely in Namaqualand, as well as in other dry parts of S. Africa, that it must be mentioned, for it is conspicuous in this semi-desert which is so lacking in trees and tall shrubs.

It forms a graceful shrub with willowy branches and grows to a height of about 3 metres. The greenish-yellow tubular flowers are produced in sprays at the ends of the branches in spring and summer, and the leaves are simple and tapering (see plate 42).

Nymania capensis
(Meliacea)

Klapperbos, Chinese Lanterns

The decorative inflated rose-pink seedpods of this large shrub festoon the arching branches in great profusion in spring, following the deep rose, bell-like flowers, which are about 3 cm long. The pods, which have a papery quality, measure about 5 cm in length and in diameter. They hang down from between the tufts of narrow leathery leaves, flowering in winter and spring. The pods remain decorative on the bush for a long period.

Nymania grows from 1 to 2 metres in height and is found in hot, dry sandy soils in the area near the Orange River in Namaqualand, between Vioolsdrift and Pofadder, as well as in S.W.A., the Karoo and parts of the eastern Cape. It is the only species in the genus (see plate 41).

Oftia revoluta
(Myoporaceae)

Oftia

A pale green shrub that grows to 2 metres, this is densely covered with small hairy leaves that are narrow and up to 2½ cm long. The edges are indented and rolled over. The small white starry flowers are scattered between the leaves near the tops of the hairy branches and appear mainly in spring and early summer. They are followed by small, juicy berries. This shrub is found in the Kamiesberg near Leliefontein and extends northwards to Okiep. There is one other species in this genus which is common in the mountains of the S. W. Cape.

Kliphout *Ozoroa dispar (Heeria dispar)*
(Anacardiaceae)

A shrub that grows to about 1½ metres, this has light green oval leaves with prominent veins, about 5 cm long and fairly broad. The small white flowers are followed by black berries on the female plant and the male flowers occur on a separate bush. The berries were noted to be poisonous to man, but not to birds, in van der Stel's journal.

This shrub occurs on sandy soil in the Kamiesberg and Spektakel mountains, but it is only noteworthy because of the few large shrubs in Namaqualand.

Passerina *Passerina glomerata*
(Thymelaeacea)

A small shrub that grows to 1 or 2 metres in height, this is much-branched near the top. The slender branches are woolly when young and covered with small narrow leaves. The tiny dark red flowers, borne in a terminal spike, are flask-shaped. This little ericoid shrub grows wild near Garies and Kamieskroon.

Polygala, Bloukappies *Polygala*
(Polygalaceae)

There are many Polygalas in South Africa, which may all be recognised by the small crest or hairy tuft at the top of the lower keeled petal. There are 3 to 5 petals and the two outer petals enclose the lowest petal like a little bonnet. Some are bushes and others are small perennials.

P. affinis is a very small shrublet with tiny needle-like leaves and typical purple flowers. It occurs in the Kamaggas mountains in Namaqualand and in the S. W. Cape.

Protea, Sugarbush *Protea*
(Proteaceae)

Of the 150 species of Protea that occur in Africa, most of which are massed in the S. W. Cape, only one, *Protea sulphurea*, is found growing wild in Namaqualand. See *Leucospermum alpinum*, which is the only other member of the Protea family to have been collected or seen in Namaqualand.

Protea sulphurea, the Sulphur-coloured Protea, is rare wherever it occurs in the drier districts of the S. W. Cape, near Robertson and Montague in the Langeberg Mountains, near Villiersdorp further south and as far north as the Swartberg Mountains on the edge of the Karoo. It has been found growing at high altitudes in the Kamiesberg mountains of Namaqualand, but is not common there either.

P. sulphurea is a large spreading shrub with long branches that may cover 3 metres in area, but it only grows to a height of about 60 cm. The flower-heads hang downwards facing the ground and reveal only the reverse side of the overlapping bracts which are sulphur-yellow and neatly edged with rose. The whole head measures about 8 or 9 cm across (see "Proteas for Pleasure").

The curving bracts open wide to reveal yellow bases, shading to rose at the tips, and a beehive-shaped mass of hairy flowers at the centre, which are sulphur-yellow and tipped with golden-tan. They have a strong scent of honey. The small tough, bluish-green leaves are broader near the tips.

Pteronia Renosterbos
(Compositae) (Rhino-bush)

Numerous species of these small, branched shrubs are common in Namaqualand, where they form an important component amongst the small bushes that cover large areas of ground. When they bloom in springtime, some are attractive.

P. incana is probably the showiest, with its bright yellow flowers, about 2 cm long, that become hairy as they mature. It has been cultivated as an ornamental. The thin wiry branches are covered with short, needle-like leaves. It occurs near Grootvlei and Kamieskroon, as well as around Springbok.

P. divaricata is similar, but has round leaves. It is found near Kamieskroon, Hondeklip Bay and in the Richtersveld.

P. undulata, a leafy shrub that grows to almost 2 metres, has creamy-yellow flowers. It grows wild near Garies, Kamieskroon and Klipfontein.

Rhus Wild Currant,
(Anacardiaceae) Karee

Over 60 species of these evergreen trees are indigenous to S. Africa and widely distributed. Two shrubby species occur in Namaqualand and may be recognised by their small leaves that are divided into 3 leaflets springing from the end of a short leaf-stalk. They bear small round fruits and the flowers are insignificant.

R. undulata var. *tricrenata* is a rounded, woody, much-branched bush of over 1 metre, which frequently acts as a host to the parasite *Loranthus*. The tiny heart-shaped leaflets are joined at the narrow end. It is found on the hills around Springbok, on the Hondeklip road and in the Cape Province.

R. incisa grows to 1 metre and is found near Steinkopf. The small oval leaflets are deeply indented and prominently veined.

Salsola zeyheri Blomkool Ganna
(Chenopodiaceae) (Kanna)

A small, many-branched shrublet, this grows to 60 cm and spreads to almost as much in width. It has small, grey-green leaves, arranged alternately on the stems, and the small, yellow flowers are covered with very woolly bracts. This drought-resisitant plant occurs between Pella and Concordia and is used as a fodder plant.

It should not be confused with Kanna or Kougoed (*Sceletium tortuosum*).

Beach Salvia, Strandsalie, Golden Salvia, Geelblomsalie	*Salvia africana-lutea (S. aurea)*

Salvia africana-lutea (S. aurea)
(Labiatae)

A sturdy aromatic shrub that may grow to 2 metres, this is frequent on coastal sandy dunes, but is also found further inland along the Namaqualand coast and in the S. W. Cape. The flowers themselves are a burnt orange or rusty-gold that is not very showy, but they are quite pleasing against the sage-like grey foliage, flowering from early spring to early summer. They also vary in colour from a brick to dull red and measure about 4 cm in length. There is a hooded upper lip and a shorter, drooping lower lip. The purplish calyx persists on the bush when the flower drops. The soft oval leaves vary considerably, but grow up to 3,5 cm long. It was illustrated by Commelin in 1701 as growing in the garden of medicinal plants in Amsterdam, according to "Flowering Plants of Africa", plate 1461.

S. dentata (S. crispula). An aromatic herb growing to 1½ metres, this has large, dark blue flowers and tiny curly, toothed leaves. It blooms in early summer and occurs from Garies to Leliefontein in the Kamiesberg.

Desert Broom — *Sisyndite spartea*
(Zygophyllaceae)

A curious shrub that would scarcely merit a second glance, this is conspicuous because it exists in a barren landscape, almost as bare as a dust-bowl. It lines the roadsides to the east of Springbok, in the driest of conditions, and also extends northwards to S. W. Africa. It is the only species in the genus.

This is a branched shrub, growing to about 1 metre in height, with broom-like rushy stems that are grey-green and almost leafless, but there are tiny oval leaves scattered along the branches. The large yellow flowers have 5 oblong petals, each about 3 cm long, forming a cup shape. The seedpod is covered with long, fluffy, creamy hairs, forming a tuft.

Bitter Apple, Giftappel, Apple-of-Sodom — *Solanum*
(Solanaceae)

The bright tomato-like berries on these plants generally catch the eye, for, although the pale mauve flowers are quite pretty, they are too common and weedy to be admired. This is a huge group of plants, but only 3 occur in Namaqualand, one of which has no thorns.

S. coccineum. A very spiny shrub from the Richtersveld, this grows to 60 cm, bearing small mauve flowers and tiny red berries about 1 cm in diameter. The oval, simple

PLATE 49

1 There are many kinds of *Aloe* in Namaqualand, which resist drought on account of their succulent leaves. *Aloe claviflora* has almost horizontal flower-spikes.

2/3 *A. krapohliana* is a small type from the Orange River region and the broad-headed Sand Aloe, *A. arenicola*, occurs along the coastal belt.

2 3

← PLATE 50

The tall Quiver-Tree or Kokerboom, *Aloe dichotoma*, was the first *Aloe* to be noticed by explorers in Namaqualand and the Bushmen used its hollow branches as quivers for their arrows. It is the most significant and dramatic landmark on the hills around Springbok, where golden Early Morning Daisies (*Osteospermum hyoseroides*) enliven the scene in the spring.

PLATE 51

1 The trunk of the Kokerboom (*Aloe dichotoma*) is rugged when mature, but satin-smooth when young.

2 The Kokerboom's gaunt silhouette is accentuated by a barren, boulder-strewn landscape, where it grows in company with other distinctive succulents, like the Halfmens, *Pachypodium namaquanum*.

1

2

leaves are up to 4 cm long, borne on hairy branches, with small spines almost half a centimetre in length.

S. gifbergense. This very spiny shrub is common from Garies through Springbok to the Richtersveld. The rough, oval leaves are lobed and up to 4 cm long, while the long reddish or yellow spines are over 1 cm in length. The small purple flowers are followed by orange berries, about 1 cm across.

S. guinense. There are no thorns on this shrub, that grows to well over a metre in height. It may sprawl and bears smooth, fleshy, simple leaves that are up to 5 cm long. The large lilac flowers are followed by yellow or orange berries, about 1 cm in diameter. This species is found near Garies and Hondeklip Bay.

Spergularia marginata — Spergularia
(Caryophyllaceae)

A bushy shrublet that grows to 30 cm, this has thin, needle-like, sticky-hairy leaves. The small white or creamy flowers have 5 petals. These little bushes are common around Springbok, Port Nolloth and southwards to Garies.

Struthiola leptantha — Katstert Cat's Tail,
(Thymelaeaceae)

An erect heath-like shrublet growing to about 30 cm, this is covered with thin, short, needle-like leaves. The tiny white flowers are crowded in spikes near the tips of the stems, each consisting of a very thin tube that is about 2 cm long, opening into 4 tiny pointed petals. It is found on the hills of Namaqualand, near Concordia, Spektakel Pass and Leliefontein, where there is a much larger-flowered form than that which occurs in the S. W. Cape. Numerous other species grow in the Cape (see plate 22).

Sutera — Sutera, Wild Phlox
(Scrophulariaceae)

Annual *Sutera* plants will be found in the section on Annuals and Perennials.

S. fruticosa. A small rounded bush, growing to about 40 cm in height, this has clusters of flowers near the tops of the stems. They are thin-tubed, opening into 5 mauve lobes with dark purple markings at the centre. They may be various shades of blue with a yellow centre and dark spots. The pointed leaves are aromatic or pungent, varying in width and slightly hairy. This species occurs in the Kamiesberg, near Springbok, Port

PLATE 52

A regiment of *Conicosia pugioniformis*, commonly called Varkwortel or Pigsroot, emblazons a vast field. The huge flowers glitter in the sunshine and attract pollinating flies and midges. Their soft, fleshy leaves store moisture so that they are able to survive extreme drought.

Nolloth and in the Richtersveld. It also grows wild in S. W. Africa and near Van Rhynsdorp in the Cape.

S. ramosissima. A densely branched bush about 1 metre in height, this spreads to the same width and is so showy in bloom that it can be seen from afar, nestling against the rocky mountain-slopes north and west of Springbok, as far as Agenneys and Pella. It is covered with small china-blue, 5-petalled flowers, each about 1 cm across, but so numerous as to obscure the foliage. The greyish-green, triangular leaves are very tiny and, unfortunately, have a most pungent and overpowering smell (see plate 45).

Balloon-pea, Gansies, Kankerbos — *Sutherlandia frutescens*

(Leguminosae)

This shrubby plant grows to a height of nearly 2 metres in favourable situations, but is half the size or less when it grows in stony clefts in the mountains of Namaqualand, where it occurs from Garies to the Kamiesberg and in the hills around Springbok. It is also widespread in the Karoo, where it is regarded as a valuable fodder plant, in S. W. Africa, N. Cape and Botswana. An infusion of the leaves was used for various ailments by tribesmen and the early colonists.

The scarlet pea flowers grow to about 3 cm long and are quickly followed in spring by large, green inflated pods, from 4 to 5 cm long and half as wide. The leaves are divided into leaflets that may reach 3 cm in length and 1 cm in width (see plate 43).

S. microphylla is a wide-spread species with much smaller, narrower leaflets, smaller flowers and seedpods.

Dawip, Dawib, Tamarisk — *Tamarix usneoides*

(Tamaricaceae)

Tamarisks are always drought-resistant and this species, which is the only one that grows wild in Namaqualand, is no exception. The old vernacular Hottentot name for it, DAWIEP, is preserved in a place name in Namaqualand.

This deciduous tree grows to 6 or 7 metres and has minute, scale-like leaves. The tiny flowers are a dirty white and grow in dense sprays. It thrives in brack soil and is common in the Buffels River bed. It also occurs in the Richtersveld and near the Orange River, extending to Upington in the northern Cape.

Klipkersie — *Teedia lucida*

(Scrophulariaceae)

A soft, prostrate shrublet with pink or mauve flowers, arranged in broad heads at the tips of the branches, this blooms in September. Each long-tubed flower, about 4 cm in length, opens into 5 lobes, and is about 1½ cm across at the top. The flowers are followed by pea-sized berries. The leaves may vary in size according to the area in which they are found and grow up to 12 cm in length where rainfall is good, but the

Namaqualand forms are smaller. The leaves are oval, with serrated edges, and are arranged in pairs alternately at right angles to the square stems and opposite one another.

Teedia lucida occurs from low to high altitudes in all provinces of S. Africa, where it may be found in rocky places in veld and mountain. It has been collected at Kamieskroon and Tweefontein in Namaqualand.

Zygophyllum spinosum Spekbossie
(Zygophyllaceae)

A small evergreen shrublet that grows to about 30 cm, this has small fleshy pointed leaves and tiny prickles on the stems. The flowers, about 2½ cm across, have 5 yellow petals streaked with red near the base and open wide to reveal the central cluster of stamens. This is found on sandy soil on the road to Hondeklip Bay, as well as in all parts of the S. W. Cape.

THE SHOWY
SUCCULENTS

The Showy Succulents

Namaqualand has long been considered the happy hunting ground of succulent collectors, for its dry climate has given rise to a race of plants that has adapted to life in a semi-desert.

The rainfall is often so negligible that the nightly dews or nearby ocean mists provide them with the little extra moisture that is vital for their existence and help them to withstand the burning sun during summer, when the temperature soars to 38°C (100°F) or more, without the relief of rain. Some find shelter in the shade provided by rocks, which also provide them with extra moisture through condensation.

The remarkable stone plants (*Lithops*), that nestle among the stones and are scarcely distinguishable from them, as well as many similarly formed dwarf plants, are to be found in their thousands. Some of the more common succulents are found both in Namaqualand and in the Karoo, while many are endemic only to these separate areas, with some rarer kinds being isolated in only one small portion of the countryside. This Karoid climate extends northwards into South West Africa and all of these dry, semi-desert areas are rich in succulents that bloom briefly in order to set seed and propagate themselves in their harsh surroundings.

As interesting as these are, however, it is not the curiously formed small plants that play a part in the floral scene in Namaqualand. Their flowers, although brilliant and colourful in themselves, emerge only in ones and twos from between the folds of the small, fleshy shapes and are too evanescent to contribute to the greater display made by the larger, succulent forms.

The showy succulents, that were formerly all called *Mesembs*, augment the floral pageant to the greatest degree. Most of these have been renamed and placed in different genera of the family *Mesembryanthemaceae*. They include *Lampranthus* and *Drosanthemum*, to mention the most attractive in the group. These form jewelled cushions of winking flowers in the brilliant spring sunshine, sometimes studding the sand with purple or gleaming from between black stones on a hillside in orchid pink, ginger or tangerine. Some of these form mounds of colour among annuals like *Ursinias*, together weaving a striking tapestry of strong colour to dazzle the onlooker.

Most succulents are perennial in habit, but some are annual, seeding themselves anew each year. The showy annual *Dorotheanthus* is classified with the succulent tribe, and, although a succulent enthusiast would not look upon it as a plant to be collected and would rank it among the showy annuals, it should be placed in its correct category among the succulents.

One of the most striking genera among the succulents is that of the *Aloe*, which is represented in all parts of South Africa as well as in tropical Africa and Socotra. Some of the members of this huge genus occur in mountains and areas of high rainfall, but most of them find a home in the dry parts of the country, flourishing in places with high temperatures in summer and bitter cold in winter, though this is seldom accompanied by killing frost. The *Aloe* is well represented in Namaqualand, varying in form from the tiny *A. variegata* to the tall, tree-like *Aloe dichotoma*. This is known as the Kokerboom or Quiver-tree, because the Bushmen made their quivers from its hollowed-out branches, and it stands like a gaunt sentinal on rocky hillsides, commanding attention. If one climbs the low koppies to look down silently through the bare branches, it seems as though they have stood there for centuries, making man's struggles seem puny by comparison.

One of the curious succulents that stimulates the imagination to flights of fancy is the Halfmens (*Pachypodium namaquanum*) meaning "Half-a-man". This is an eerie plant to meet by moonlight, standing in groups like tall old men nodding in consultation. Another succulent of special interest is the Elephants Foot (*Dioscorea elephantipes*) which was almost rendered extinct at one time by collectors, not only because of its strange appearance, but because it was one of the natural sources of cortisone, which is now produced scientifically.

One cannot ignore the curious succulents that never cease to amaze and fascinate the traveller, not yet the unusual terrain in which they grow, varying from the moon-landscape of the Orange River valley to desolate hillsides or coastal sands. Most of the succulents described below are of the showier kinds that do most towards creating the flowering semi-desert of Namaqualand, but some of the peculiar or remarkable species among the smaller or singular types have also been included.

PLATE 53

The Giant Sour Fig, *Carpobrotus quadrifidus,* has immense spectacular flowers and large thick leaves. It grows wild on the sandy coastal belt.

1

2

1

- PLATE 54

1 The chief species of *Dorotheanthus* in Namaqualand, *D. bellidiformis*, is a dwarf annual Mesem. with glittering flowers, which generally have a white ring around the yellow centre. They may also be orange, white or rose.

2 Olifantsbos (*Euphorbia dregeana*) is a succulent shrub with spineless, thick, cylindrical stems. It grows in very dry sandy soil and is carpeted here with blue Sporries (*Heliophila*) and yellow *Grielum*.

PLATE 55 →

1 Half-mens, meaning Half-a-Man, is the name prompted by the romantic imagination, heightened by the desolate eerie landscape near the Orange River where it is found, for the strange succulent *Pachypodium namaquanum*. It has long been a subject of legend among the Hottentots.

2 The velvety flowers emerge in August and September among the topmost leaves.

2

1

Adromischus Adromischus
(Crassulacea)

Small succulent plants from the Cape and Namaqualand, these have rounded, fleshy leaves and are often stemless. There are 28 species.

A. herrei is said to be the most notable of the Namaqualand species, with a cluster of distinctive leaves that are rough and scaly on the surface. The flower-stem grows up from the centre, bearing several green, five-pointed flowers that are tinged with red and grey. This species comes from the Richtersveld.

Aloe Aloe,
(Liliaceae) Aalwyn

Aloes are adapted to withstand drought and one would expect to find them well represented in Namaqualand, but only 16 of the 132 South African species grow wild in this area. These vary from dwarf plants like *A. variegata* to the tall tree Aloe, *A. dichotoma,* that is known as the Kokerboom and forms such a dominant feature on the hills around Springbok.

As Aloes are so well-known and popular among gardeners, all the species that are found in Namaqualand will be mentioned here.

A. arenicola. SAND ALOE.

Formerly plentiful along the sandy coastline near Port Nolloth and Hondeklip Bay, extending southwards to Lambert's Bay and northwards in S. W. Africa, these have been reduced in number, chiefly by grazing goats, as well as by collectors. This is a short plant, growing to 75 cm, with a rosette of bluish-green leaves at ground level. These are white-spotted, with horny white margins and teeth. The curving stems sprawl and turn upwards as they mature and are clothed with the leaves pointing upwards along their length. The flowers are pinkish or orange-red and form large rounded heads at the tips of the stalks, blooming in early spring or in summer (see plate 49).

A. claviflora. KRAAL AALWYN (KRAAL-ALOE), KANON AALWYN (CANNON-ALOE).

The striking flowers on this species emerge at the sides of the cabbage-like clusters of greyish-green leaves, lying almost horizontally on the sandy soil. The young flowers and buds are bright reddish-brown and turn yellow or cream as they mature at the base of the long spike. The leaves are curved and pointed, forming an upright rosette of 20 to 25 cm in height, with a few teeth near the top of the keel. The plants multiply

PLATE 56

1 A moisture-loving dwarf succulent, *Crassula natans,* forms crimson patches in shallow depressions wherever water collects. The tiny mauve flowers appear during spring.

2 The curious Elephant's Foot, *Dioscorea elephantipes,* has a woody enlarged stem that resembles an elephant's foot. A climbing stem arises from the centre.

and grow closely together in masses. This species blooms in August and September. This is very wide-spread in the Karoo and northern Cape, especially in the dry portions near the Orange River. It is found in the Richtersveld in Namaqualand and extends into S. W. Africa (see plate 49).

A. *dichotoma*. KOKERBOOM, QUIVER-TREE.

Colonies of these tall, tree-like Aloes create a dramatic silhouette on the crests of hills to the north, east and west of Springbok, where they are common, but they also occur as far south as Garies and northwards into S. W. Africa. They are generally seen on the hot northern slopes of the hills, but also grow amongst boulders in the most arid portions of the country, especially near the Orange River.

The Kokerboom, meaning "Quiver-tree", owes its common name to the fact that Bushmen used the hollow branches as quivers in which to carry their arrows. The tree may grow to 4½ metres in height and becomes gnarled with age. It is satiny and smooth when young. The branches fork and rebranch to form a rounded crown, bearing large rosettes of leaves at the ends. The flower-spikes emerge like erect yellow candles from the centre of the leaves and bloom during winter in June or July. The flowering period is over when the hillsides come alive with yellow daisies in springtime, forming a carpet between these gaunt sentinels (see plates 50, 51).

A. *falcata*.

A medium-sized plant that grows to 75 cm, this has large rosettes of sickle-shaped leaves that spread over the sandy soil. They are greyish-green, spine-tipped and edged with sharp, reddish-brown teeth. The flower-stem is branched and rebranched to form a candelabra of 9 or more pointed flower-spikes, which may be pale scarlet or orange. This plant blooms during December and is common in local areas throughout Namaqualand in dry sandy country, from the Orange River southwards to Klawer in the S. W. Cape.

A. *framesii*. FRAMES' ALOE.

Common in localised areas, this grows on sandy plains near the coast, from the Orange River southwards through Namaqualand to Saldanha Bay, Cape. This plant grows to 1 metre in height and the large rosettes spread to form large clumps on the soil. The leaves are bluish-green tinged with pink and usually spotted with white. The flowers bloom in June and July, forming a pointed spike of orange-scarlet flowers.

A. *gariepensis*. ORANGE RIVER ALOE.

This species occurs in a wide band on either side of the Orange River in the northern Cape, Namaqualand and S. W. Africa. It has a solitary rosette and nestles atop rocks, with several flower-stems arising from each rosette to a height of just over 1 metre. The pointed leaves are yellowish-green or tinged with red and are edged with reddish-brown teeth. The long, slender-pointed flower-spike may be plain yellow or have orange buds opening into greenish-yellow flowers at the base of the spike. It is found in bloom from July to September.

A. glauca. BLOUAALWYN (BLUE ALOE).

As the name indicates, this plant has bluish-green, smooth pointed leaves, often marked with thin lines. They are edged with small reddish teeth. These form a low rosette among rocks and the flower stems emerge from these to reach a height of almost 1½ metres, bearing flowers from August to October. The flower-spikes are short and thick with large, pale pink or apricot flowers. This species is common in the hills around Steinkopf and Springbok and spreads into the Karoo and towards Swellendam and Piketberg in the S. W. Cape, where it varies in size and form. This variation is thought to be a result of differing climatic conditions.

A. karasbergensis. KARASBERG ALOE.

An attractive plant with a low-growing rosette of very broad, smooth grey leaves, these are marked with lines and have no spines. The slender flower stem makes a pyramid of many branches, festooned with individual flowers that form a pale coral-red mass. It resembles *A. striata*, but has more pointed leaves and a more branched flower-stem. It is common in the area around Karasberg and near the Orange River, and is widely distributed in the N. Cape and S. W. Africa.

A. khamiesensis. KAMIESBERG ALOE.

Named after the Kamiesberg mountains of Namaqualand, the specific name is spelt with an "h", as in the older spelling of this name. This species forms clumps on the rocky hillsides of the mountains near Kamieskroon and northwards. They grow to a height of 2½ metres when mature. The rosette of narrow green leaves is at the top of a thick stem that is covered with the dry remnants of old leaves. The flower-stem emerges from the top and branches into several stalks that bear 7 or more spikes of flowers in June and July. They are orange-scarlet and form a cone-shape that is pointed at the tip.

A. krapohliana. KRAPOHL'S ALOE.

A small Aloe, this has short, rounded rosettes of bluish-grey, curved leaves at ground level, sometimes branching into two or three rosettes clumped together. They are sometimes marked with dark bands across the leaves and the edges have short white teeth. The flower-stems emerge singly and bloom from mid-winter to early spring, in June, July and August, reaching a height of about 40 cm. The broad, tapering flower-spike has large rosy-scarlet or scarlet flowers. This species is common near the Orange River, near Steinkopf and eastwards to Pella, extending in scattered areas southwards to Van Rhynsdorp (see plate 49).

A. melanacantha. GOREE, KLEINBERGAALWYN.

A rough-looking plant, this forms colonies of small rounded rosettes spreading over the rocky or sandy soil in arid areas. The narrow leaves are a dull brownish-green, with black thorns on the teeth and margins. The name *melanocantha* means "black thorns". Flower-stems, growing to 1 metre in height, are simple, bearing dense, pointed spikes of pinkish-red or orange flowers in June. This Aloe is common on slopes and hills throughout Namaqualand, extending as far south as Bitterfontein and northwards into S. W. Africa.

A. mitriformis. MITRE ALOE, KRANSAALWYN.

This species is closely related to *A. arenicola* and grows in the same region, but is found further inland, perching on rocks instead of on the sands. The leaves of *A. mitriformis* are broader and sparsely spotted, if at all, in contrast to those of *A. arenicola.* (See Flowering Plants," plate 1467). The specific name refers to the appearance of the topmost leaves, which resemble a bishop's mitre.

A. pearsonii. PEARSON'S ALOE.

Radically different in appearance from all the species described above, this plant forms a bushy shrub of about $1\frac{1}{3}$ metres in height. It is composed of several thick stems, covered with broad, pointed, grooved leaves, overlapping downwards as they ascend. The flower-stems emerge at the tops of the stems during mid-summer, bearing broad, pointed spikes of yellow flowers. There is a brick-red colour form. This species occurs among stony hills in the Richtersveld and the Orange River valley in northern Namaqualand, extending into S. W. Africa.

A. pillansii. PILLANS' ALOE.

A tall tree Aloe, this is as tall as the Kokerboom, but differs from it in its comparatively sparse branches springing from a shorter, broader trunk. They are more erect and bear a greyish or brownish-green leaf rosette at the top of each branch, with the golden yellow flowers hanging downwards in September and October in a fringe below the leaves, unlike those of *A. dichotoma*, which stand erect. *A. pillansii* is not seen very often, as it occurs in the Richtersveld near the Orange River, as well as in S. W. Africa, in rocky, desolate and arid places.

A. ramosissima. BRANCHING ALOE.

As the name indicates, this is a many-branched bushy Aloe, which forms a shrub of about 2 to 3 metres in height. The yellowish-green leaves, sometimes red-tinged, are concentrated near the tips of the stems, forming lax rosettes, and the thick spikes of bright yellow flowers emerge from between them in June and July. This species grows in rocky, arid places in northern Namaqualand near the Orange River, as well as in S. W. Africa.

A. variegata. KANNIEDOOD, VARIEGATED ALOE.

This charming small Aloe is a great favourite because of its attractive thick, smooth, dark green, thornless leaves that are speckled and edged with white. They form a small neat rosette on the ground, no taller than about 12 cm, with short stems bearing lax spikes of pale red flowers emerging in September. This Aloe is wide-spread in the Karoo, overlapping into the Cape, the O.F.S. and S. W. Africa, as well as into Namaqualand. It is generally found in open, dry soil, often sheltering from the hot sun in the shade under small bushes or amongst stones.

Anacampseros alstonii
(Portulacaceae)

Paper-plant,
Haasieskos,
(Rabbits-food),
Moerplant

The species with papery stems are the most interesting in this large genus. Several of these occur in Namaqualand and *A. alstonii* is possibly the best-known.

This is a dwarf plant with branching stems only a few centimetres in height. The stems are completely covered with overlapping silvery, papery scales. The flowers are fleeting and insignificant. This curious little plant is found growing in quartzite gravel in the shade of rocks in northern Namaqualand, near Pofadder.

Apatesia helianthoides
(Mesembryanthemaceae)

Apatesia

An attractive annual plant, this has silky yellow daisy-like flowers borne singly at the top of thin stems, arising from a tuft of succulent leaves near the base. The flowers measure about 3 cm across and are composed of 4 or 5 rows of slender petals laid one above the other. The broad centre is filled with numerous stamens that fill a saucer-shaped calyx. The spoon-shaped, fleshy, pale green leaves, tinged with red, are 3 – 4 cm long and taper at both ends. The whole plant grows to a height of 15 to 20 cm.

Apatesia helianthoides grows in sandy soil in the area around Springbok and southwards to Kamieskroon. Three other species occur in the S. W. Cape.

Astridia
(Mesembryanthemaceae)

Astridia

There are 10 species of *Astridia* which occur in the Orange River valley in the Richtersveld, extending northwards into S. W. Africa. They form erect shrubs growing to 75 cm, with long thick, succulent leaves, sometimes as long as 11 cm and as wide as 3 cm, so that they are among the largest in the family. The flowers are 5 cm in diameter and may be red, pink, rose or white. The stamens form a cone in the centre and the petals turn up to form a saucer-like shape. *A. longifolia*, with red flowers, one of the brightest in the genus, occurs in the Richtersveld, Namaqualand.

Berrisfordia khamiesbergensis
(Mesembryanthemaceae)

Berrisfordia

This tiny perennial has thick leaves with a warty or knobbly texture, clustering at ground level to form a small cushion. They have a cleft at the top and the pink flowers emerge from the centre in springtime. Each daisy-like flower is about 3 – 4 cm across. This is the only species in the genus and it is found in the Kamiesberg near Garies in Namaqualand.

Sour Fig, Gouna, Gaukum   *Carpobrotus*

(Mesembryanthemaceae)

These perennial succulents have large edible juicy fruits, called Sour Figs. They may be recognised by their large, fleshy leaves that have triangular sides and taper to the tip. These are numerous on the long trailing stems that lie on sandy soils, particularly near the coast. There are 20 species in S. Africa that occur along the Cape coast, from Namaqualand to Humansdorp in the east.

C. quadrifidus. GIANT SOUR FIG.

This species has huge flowers, from 10 to 15 cm across, which may be plain white, becoming pink with age, or a brilliant cerise, with a white band around the centre. The large leaves are about 10 cm long and $2\frac{1}{2}$ cm thick. This spreading plant is found along the coast of Namaqualand and extends southwards to Calvinia (see plate 53).

C. edulis and *C. muirii* are well-known Cape species which do not occur naturally in Namaqualand.

Cluster-leaved Mesemb   *Cephalophyllum*

(Mesembryanthemaceae)

These are not very large plants as their stems are fairly short and lie on the surface of the soil. They may be recognised by the ball-like tufts of thick leaves placed at intervals along the stems. These leaves are triangular, like those of *Carpobrotus*, but are not so sharply angled and are much shorter.

C. spongiosum. This is one of the most beautiful in the genus, with rich deep pink flowers, about 5 cm in diameter, which emerge on stalks from the leaf clusters in spring-time. The bluish-green leaves are larger than most, up to about 10 cm long.

This is found in Namaqualand and there are 63 other species, which occur in Namaqualand, especially in the Richtersveld, the Cape, Karoo and S. W. Africa. Their flowers may be yellow, salmon, pink or red.

Cheiridopsis   *Cheiridopsis*

(Mesembryanthemaceae)

These dwarf, succulent perennials have large yellow, orange or white flowers, measuring about 4 – 10 cm across. These nestle above clumps of long, bluish-green leaves, which are generally heavily-dotted, boat-shaped and sharply keeled, about 10 cm long and of unequal length.

C. tuberculata is the type specimen, which is found in Namaqualand, especially in the Richtersveld, in S. W. Africa and the S. W. Cape. It was described as long ago as 1768 in the Gardener's Diary as *Mesembryanthemum tuberculatum*. It has golden-yellow flowers.

C. candidissima is an even more robust common species, with flowers that may be white or yellow. It grows in sandy soil in northern Namaqualand.

C. peculiaris is an interesting species from the Steinkopf area. It has only 2 broad leaves, about 10 cm long and 6 cm wide. These lie flat on the soil during autumn and winter and are covered with a protective thin membrane during the hot summer months, when they almost disappear from sight. The flower is bright yellow.

Conicosia pugioniformis
(Mesembryanthemaceae)
Varkwortel, Pigsroot

Striking large silvery-yellow flowers, fully 8 to 13 cm across and shimmering like tinsel in bright sunshine, make an unforgettable sight. They have a double row of thin petals and resemble a Chrysanthemum. The solitary flowers stand well above tufts of long, pencil-shaped green leaves, which have an extremely fleshy and sappy texture. A single plant is decorative, festooned with several large flowers, whilst a whole field of them, with all the flowers facing towards the sun, is particularly arresting. They open fully around midday and remain open until about five in the afternoon, attracting myriads of small black midges and buzzing flies with iridescent green eyes, which pollinate them.

These Conicosias are dotted individually on the dry sandy soil, even though they may grow in huge drifts covering large areas. Enormous fields of *C. pugioniformis* may be seen at Sandhoogte, a few miles west of Springbok on the road to Spektakel Pass, as well as east of Springbok, half-way to Pofadder. It also occurs in the S. W. Cape. There are 9 other species in the genus, which are very similar. Conicosias are perennial plants with fibrous or fleshy rootstocks and are extremely drought-resistant (see plate 52).

Cotyledon
(Crassulaceae)
Cotyledon, Pig's Ear, Varkoor, Hondeoor

There are about 50 South African species in this large group of succulents, that is represented in many other countries. Several of these are favourite garden plants, such as *C. orbiculata* and *C. mucronata*, but neither grows wild in Namaqualand.

C. decussata. BERGBESIE.

This succulent branched shrublet has clusters of pencil-shaped leaves that grow to 12 cm in length and taper to a point at the tip. They reach a height of about 40 cm and the long flower-stems shoot up from between the leaves, bearing clusters of drooping, pinkish-red flowers that are tubular and open into spreading lobes. This species grows in rocky places in Namaqualand, as well as in the O.F.S. and the Cape Province.

C. paniculata. BOTTERBOOM (BUTTER-TREE).

A curious plant that grows to almost 2 metres in height, this has a thick robust trunk that branches near the top. It is covered with a yellowish papery bark that peels off with age. The shiny, fleshy green leaves are clustered near the ends of the branches and drop off before the flowers develop. The barren trunks have a grotesque and interesting appearance when they are seen growing on rocky hillsides. The flowers appear after most Karoo bushes have flowered, in late spring or summer, at the ends of thin stalks that branch in all directions. Each dark red tubular flower is streaked with yellowish-green and opens into recurved segments.

This species grows wild in the dry areas of the Karoo and S. W. Africa. It occurs near Vioolsdrift in Namaqualand.

Crassula *Crassula*
(Crassulaceae)

Many popular garden plants belong to this large group of over 300 species, several of which occur in Namaqualand. They vary greatly in appearance, but generally store water in their succulent leaves and stems.

C. columnaris. Quaint miniature plants that are collector's favourites, these grow to a height of 15 cm. Thick oval leaves curve inwards in 4 closely set rows, overlapping one another to form a thick column. A rounded head of tiny greenish-white flowers develops at the top in early summer. These little plants are found growing in the shade among rocks in the Kamiesberg and in the Richtersveld, as well as in the Karoo and S. W. Cape. This species should be compared to *C. pyramidalis.*

C. expansa. BOSDUIFGRAS, BUSH-DOVE-GRASS.

This dwarf perennial has short grassy stems with tiny, oval green leaves that are sometimes reddish, and tiny flowers. It grows in the dry veld around Garies as well as in the S. W. Cape. It is not a water-loving plant like *C. natans.*

C. incana. A woody shrublet that grows to a height of 30 cm, this has tiny fleshy leaves at intervals up the stem. There are small crowded heads of tiny red flowers at the tips of the stems. This species grows wild in the Karoo and Namaqualand.

C. namaquensis, a dwarf species from the Port Nolloth area, has insignificant whitish-green flowers, unlikely to attract average attention. There are several other similar dwarf species.

C. natans. This interesting plant is found in shallow saucer-like depressions in the soil where water collects. Its specific name, "natans", means "swimming" and this is a moisture-loving plantlet that forms a veil of soft, waxy, scarlet stems on the ground, with tiny leaves. When one drives past swiftly, it appears as a patch of crimson on the soil. Tiny mauve flowers appear in spring, resembling *Alyssum.* This dwarf groundcover is common around Springbok and Kamieskroon, as well as southwards to Vredendal in the S. W. Cape (see plate 56).

C. pyramidalis. Another collector's favourite, this dwarf succulent grows to a height of about 7 cm and is composed of 4 rows of thick triangular green leaves, pressed closely on top of one another to form a column. The tiny white flowers form a head at the top of the stalk and bloom in springtime. The plant generally dies after the seed has ripened. This species occurs in the Karoo as well as in Namaqualand.

C. tomentosa. This small shrublet has small broad leaves in a rosette near the base and a finger-thick stalk covered with whitish felt. There is a long spike of tiny pink flowers arranged in whorls at intervals up the stems. This species is found near the coast at Soebatsfontein, as well as in S. W. Africa.

Dorotheanthus bellidiformis Bokbaai Vygie,
(Mesembryanthemaceae) Livingstone Daisy

These attractive little annuals form small individual rosettes of oval tapering leaves that sprawl on the ground. These grow up to 6 or 7 cm in length and are covered with shining, glassy, raised dots, which are also present on the stems. The leaves are usually green, but may also be tinged with red.

The flowers measure just over 4 cm across and are borne singly on long stalks. They have 2 rows of glistening narrow petals, which may be a clear rose, cerise or pale pink. They may also be orange, plain white or white, heavily tipped with red, so that they appear to be red with a white ring around the dark centre. They hybridise to produce subtle shades of salmon, beige or rose with a broad band of white or pale colour around the centre.

This is the chief species of *Dorotheanthus* in Namaqualand, where it may be seen near Springbok, as well as on the sandy plains well to the west of Springbok and in the Soebatsfontein area. *D. bellidiformis* also occurs near the coast in the S. W. Cape near Darling and Ysterfontein. There are 10 other species in the genus (see plate 54).

Drosanthemum Dew-flower
(Mesembryanthemaceae)

The Dew-flowers are so-called because they have glistening, glassy raised dots on their leaves, sprinkled like icing-sugar over the green surfaces. The open flowers are generally brilliant, seen at their best in strong midday sunshine. There are over 95 species in the large genus and they occur in the drier portions of the S. W. Cape, the Karoo, Namaqualand and S. W. Africa.

D. speciosum. Possibly the most beautiful in the whole group, this forms a spreading bush of about 30 cm in height, which is covered during springtime with glowing orange flowers, fading to white at the centre, or brilliant crimson fading to beige at the centre. Each flower measures about 5 inches across and has one or two rows of narrow petals. They stand well above the foliage, each on a wiry reddish stem. The small, cylindrical leaves are lightly covered with "icing sugar" and arranged in pairs at intervals along the thin, wiry stems (see plate 48).

D. hispidum. Deep mauve flowers characterise this species and they form glittering small flat cushions on the sandy soil in Namaqualand, as well as in the Karoo. The light green leaves are thickly covered with "icing sugar" (see plate 48).

Euphorbia Euphorbia,
(Euphorbiaceae) Melkbos

This enormous group of succulents, said to consist of over 1 000 species distributed throughout the world, is well represented in Southern Africa. Leathery in texture and exuding a milky latex if broken, they are adapted to withstand dry conditions and form a feature of semi-desert localities such as are found in Namaqualand. Their curious shapes are more interesting then their flowers. Enthusiasts should study these in "The Succulent Euphorbieae", by White, Dyer and Sloane.

Euphorbia dregeana. OLIFANTSBOS (ELEPHANT'S BUSH).

This is a spineless succulent shrub, from 1 to 2 metres in height, that forms a rounded shape. Numerious cylindrical thick stems radiate from the base, each about 2 to 5 cm in diameter. They are whitish-green and fairly smooth, as the tiny leaves appear only on young growth and soon drop. The flowering stems are much thinner, about 1½ cm thick, bearing the small flowers near the tips. This rugged bush seems strangely out of character when dainty annuals like *Heliophila* cover the red soil between them on the sandy plans west of Springbok.

E. dregeana also occurs near Steinkopf and between the Buffels and Orange Rivers, as well as in the N. Cape and S. W. Africa (see plate 54).

E. gummifera. BLOUMELKBOS.

This is very similar to the green-stemmed *E. mauritanica*. It is a spineless shrub that branches freely at the base to form large clumps about 1 metre in height. The branches are about 5 to 10 mm thick and slightly ribbed. The tiny leaves appear on the young growth and drop rapidly. The branches are covered with a gummy substance that protects the plants from gritty sands and prevents transpiration. It is odoriferous and not grazed by animals. This species occurs near the Orange River and in S.W.A.

E. mauritanica. YELLOW OR GREEN MELKBOS.

This is one of the most common species. It occurs in all parts of the Cape Province as well as in Namaqualand, where it is seen on sandy soil around Springbok. It grows to a height of about 2 metres and has pencil-thin, many-branched, light green stems that form a rounded bush. The stems are smooth and spineless and the small leaves that appear on the young growth soon drop off. The yellow flowers, that appear in clusters at the top of the stems in early spring, persist for a long period.

E. stellaespina. STARRY-SPINED EUPHORBIA.

This species forms dense clumps of branches in the shape of a low shrub growing to 45 cm. Each branch is from 3 to 7,5 cm thick, with 10 to 16 angles. It is covered with stout, short spines that branch into 3 to 5 rigid, sharp spines, forming "stars". This species grows near Springbok, as well as in other parts of Namaqualand, the Karoo and the northern Cape. It was discovered during Simon van der Stel's expedition and drawn by Claudius.

E. virosa.

A striking plant, this species is conspicuous in the barren country around the Orange River, near Viooolsdrift, the northern Cape and S. W. Africa. It is a spiny succulent shrub with the main stem buried in the ground or only up to 30 cm above soil level, bearing numerous erect branches that form large clumps up to 3 m across. Each branch grows from 1 to 2 m in height and a few secondary branches appear near the top. Each branch is about 6 cm thick, with 5 to 8 angles and deeply grooved, with long spines forming a horny margin along the angles. The branches are constricted regularly at intervals of 5 to 8 cm. The highly poisonous latex was used by Bushmen in tipping their poisoned arrows.

Gasteria pillansii — Gasteria
(Liliaceae)

This succulent has the typical tongue-shaped, fleshy leaves of the genus, which are arranged in two rows. The leaves grow up to about 25 cm in height and the flower-stalk emerges from the centre, growing to almost 50 cm or more, with a long spike of typical, tubular pink and green flowers, drooping from little stems. This species is the only one known to come from Namaqualand, where it occurs in hilly country in the Richtersveld and elsewhere.

Haworthia setata — Haworthia
(Liliaceae)

A rare *Haworthia*, this is one of the very few Namaqualand species of the well-known genus that is much-loved by succulent growers. It forms a cabbage-like cluster of fleshy leaves, edged with long white bristles, which grows to a height of about 10 or 15 cm. The stem, which emerges from the centre of the leaf-cluster, bears small whitish tubular flowers. This little plant grows in rocky fissures in the hills where some moisture collects.

Hoodia — Hoodia, Wilde Gnaap
(Asclepiadaceae)

These curious succulents have thick stems, often 4 or 5 cm in diameter, that grow to a maximum height of 1 metre, forming large clumps. These have 12 to 17 angles and are edged with thorns. The large decorative flowers emerge near the tops of the stems and are shallow, almost circular and bell-shaped. They are usually pale yellow or pale pink. This genus is closely related to *Stapelia*. There are 17 species and several occur in Namaqualand.

H. gordonii. GORDON'S HOODIA, WILDE GNAAP.

This is one of the best-known in cultivation and is striking when it blooms in profusion. The plant grows to a height of about 45 cm and the flowers are pale purple with pale greenish-yellow stripes. Although hairless, it has a velvety appearance. This species may be seen in northern Namaqualand in the same areas in which *Pachypodium* is found. It occurs near the Orange River, in the northern Cape and in S. W. Africa.

Lampranthus — Lampranthus, Mesemb, Vygie
(Mesembryanthemaceae)

Undoubtedly the most showy genus in the family, as well as one of the largest, these are perennial succulents that are among the most popular garden plants and form a brilliant spectacle in the wild in springtime. They vary from spreading, creeping plants to rounded bushes of about 45 cm in height, spreading well over a metre in extent. The flowers glitter in the sunshine, opening fully only when the sun is hot and strong and closing in the late afternoon. They are so closely massed that they often cover the entire plant with colour, forming jewelled cushions on the sandy plains between other shrubs or on rocky banks.

It is not easy to distinguish between the 178 species that grow wild in the Cape Province, where these are massed mainly in the Karoo, S. W. Cape and Namaqualand. They also occur in the eastern Cape, Natal and S. W. Africa.

Most of the species of *Lampranthus* that occur in Namaqualand have purple or rose flowers. There are none with yellow or orange flowers, and only one with white flowers. One of the most common purple species, *L. zeyheri*, does not occur in this region. The following have been collected in the Namaqualand area. Further descriptions may be found in "A Handbook of Succulent Plants" by Jacobsen.

L. brachyandrus. The purplish flowers of this species are about 4½ cm across. This is a shrubby plant growing to 40 cm in height with erect branches. The leaves measure 2,5 to 3 cm in length and are about half a centimetre in thickness. This is found from Port Nolloth to Oograbies Poort.

L. comptonii var. *angustifolius*. This small bushy plant is found near Springbok and grows from 15 to 23 cm in height. It is loosely-branched, but the branches are very short. The flowers are mauve on the outer portion and white at the centre, measuring about 2½ cm across. The leaves are not very thick. The species itself does not occur in Namaqualand, but further south (see plate 47).

L. densipetalus. A species with many-petalled white flowers, this forms an erect loosely-branched shrub that grows to a height of 26 cm. The older stems spread on to the soil. The flowers measure about 3½ cm in diameter and are borne singly on stalks above the foliage. The leaves are about 2 cm long and 3 mm thick (see plate 48).

L. godmaniae. This is the most frequently seen species around Springbok, forming a small shrub up to 35 or 40 cm. The flowers are purple and about 4 cm in diameter. The branches are fairly stout and the leaves are from 3 to 4 cm in length and 5 mm in thickness.

L. plautus. A spreading shrublet with purplish flowers, this is distributed between Kamieskroon and Springbok. The flowers measure about 3½ cm in diameter. The short slender, stiff stems are 14 cm long and the branchlets are densely leafy at the tips. Each leaf is 2½ cm long and 5 mm thick.

L. suavissimus. A robust, erect shrub that grows to 1 metre in height, this has beautiful pink flowers about 5 cm in diameter. The long leaves measure from 2½ to 3½ cm and are 4 mm in thickness. This shrub grows near the coast at Hondeklip Bay.

Stone-plants, Lithops

Lithops
(Mesembryanthemaceae)

Curious little plants that lie embedded in the soil, resembling the pebbles scattered around them, these have always fascinated succulent collectors, so that they have become rare and are now protected by law.

Each little plant consists of a cylindrical fleshy body composed of two leaves joined together, but with a cleft across the centre through which the single yellow or white flower emerges in early spring, opening in the afternoon. Some have transparent tops, like tiny windows that let in the light. Half buried in the sand, as a protection from fierce sunshine, they are difficult to discover in nature. There are 79 species of *Lithops*, scattered in the drier parts of South Africa and S. W. Africa, with many of these in Namaqualand.

L. herrei, named after H. Herre of Stellenbosch, forms clumps of 10 to 15 bodies, each about 2½ cm in height and 1½ cm across, which are brownish-green in summer and greener while growing during winter. The "window" is spotted and the flowers are yellow. This species occurs near Alexander Bay in Namaqualand.

L. dinteri occurs near Viooisdrift and is more cone-shaped, but narrower near soil level and about 3 cm high. The fissure running across is very deep and the transparent window has red dots scattered on the surface, with dark dots surrounding it. The flower is yellow.

Mesembryanthemaceae — Mesemb family, Vygie, Ice-plant

This large family has been separated into about 125 genera, and only one now remains bearing the name *Mesembryanthemum*.

The enthusiastic traveller, who wishes to name these plants in the field, should consult "The Genera of the Mesembryanthemaceae" by H. Herre, for only by comparing the excellent botanical illustrations with the living plants could one hope to place them in their correct groups. Detailed descriptions alone would be impossible to follow, taking into account the bewildering number of lesser-known genera, let alone species.

There are 53 genera of *Mesembryanthemaceae* that occur in Namaqualand, varying from miniature succulents to low shrubs. Twelve have been selected for mention in this book, as they are amongst the most beautiful or distinctive and most likely to be noticed by the layman. These are *Apatesia, Astridia, Berrisfordia, Carpobrotus, Cephalophyllum, Cheiridopsis, Conicosia, Dorotheanthus, Drosanthemum, Lampranthus, Lithops* and *Mesembryanthemum*.

Mesembryanthemum — Mesembryanthemum, Vygie

(Mesembryanthemaceae)

These are very fleshy plants, covered all over with large raised glistening dots or *papillae*. The stalked leaves are long and wide, some being enormous. The flowers are arranged in clusters at the ends of the branches and are not particularly attractive.

There are 74 species in this genus, which are wide-spread in the Cape Province and S. W. Africa, as well as in the Mediterranean region, Arabia and the Atlantic Islands. Several occur in Namaqualand and the following two are most likely to be noticed by the traveller.

M. barklyi. This large-leaved species is distinctive even when it is not in flower, for it forms an erect perennial plant growing to 60 cm or more, with thick stems and branches and very large leaves. These are up to 28 cm long and 18 cm wide, very thick, wet and fleshy, with prominent nerves and covered in glassy *papillae*. The leaves are reported to have been used by the Nama people like sandpaper for removing the hairs from the hides of buck and this use was followed by the early farmers. The flowers are whitish and measure about 3½ cm wide.

This species grows on red sandy soil near the coast west of Springbok and southwards to Hondeklip Bay and the western Cape, as well as in the Richtersveld. It may also be seen between Springbok and Steinkopf.

M. macrophyllum is a prostrate plant with the largest leaves in the genus, measuring up to 40 cm in length and 32 cm in width. The flowers are pink. This occurs between Springbok and Nuwerus in Namaqualand, as well as in the S. W. Cape. It was also used by the Namas and early colonists for cleaning hides.

M. tortuosum. KOUGOED. See under *Sceletium tortuosum.*

Half-mens (Half-a-man) — *Pachypodium namaquanum*
(Apocynaceae)

A succulent shrub, this has a single swollen grey trunk, rarely branched, that grows slowly to a height of 3 or 4 metres in its natural surroundings. It is covered with spines, which are long, brown and downward-pointing in the upper half of the trunk and short near the base. The top part of the trunk inclines slightly towards the north and is crowned with a thick rosette of ruffled leaves, which drop off during the dry summer. The flowers are clustered in spring at the centre of the leaves, facing upwards. Each flower is tubular, about 4 cm long, and opens into 5 short lobes. It is velvety in texture, a light greenish-yellow and flushed with crimson at the tip and inside.

There has always been much speculation about this strange-looking succulent, which is thought to live to a great age. Some say that each row of spines represents a year of growth. Being found in desolate, boulder-strewn countryside, near the Orange River, it has been the subject of many legends among the Namas, who compare it to the spirits of men. The common name of Half-mens, meaning Half-a-Man, is based on the fact that it seems to resemble men nodding, as it were, in conversation, especially if seen by moonlight. It is a rare plant, protected in nature, and seldom seen because of the difficult terrain in which it grows. Other succulents which grow near it include Aloes, Euphorbias, *Hoodia gordonii* and many miniature succulents, as well as the beautiful Desert Rose, *Hermannia stricta* (see plates 51, 55).

Bushman's Candle — *Sarcocaulon rigidum*
(Geraniaceae)

This succulent spiny shrublet grows to a height of about 20 cm. It has a tap root and a thick, succulent, waxy stem covered with long spines, about 3 – 4 cm in length. The simple leaves are kidney-shaped and attached to the stem at their slender ends, in the axils of the spines. The large, cup-shaped flowers have 5 broad petals, each about $2\frac{1}{2}$ cm long. The plant has a black, tarry juice. This curious little plant occurs in the Richtersveld in Namaqualand.

Senecio (Kleinia) — Succulent Senecio
(Compositae)

The fleshy-leaved types of *Senecio* that were formerly called *Kleinia* are grouped under succulents as this makes them easier to identify. A few species occur in Namaqualand and, while they are not as showy as the annual species, they are worth noting because they occur in places where few other plants thrive.

S. cephalophorus. A shrubby plant that grows to 60 cm, this resembles a Cotyledon when it is not blooming. The dull, bluish-green leaves have small heads of bright yellow flowers that are clustered together in tufts. There are no ray flowers. The fleshy leaves are narrow and up to 10 cm in length. This species occurs in the Richtersveld and as far south as Springbok.

S. corymbiferous. This shrub also occurs in the Richtersveld and at Port Nolloth. It grows to 60 cm and has slender, pale green fleshy leaves up to 8 cm long. Broad clusters of tiny flower-heads stand above the foliage on long stalks, with both yellow disc and ray flowers.

S. haworthii. This bush from the Richtersveld is interesting because of its whitish effect, as all parts of it are covered with white wool. The fleshy leaves grow to 10 cm in length and the yellow flower-heads, which have no ray florets, measure 2 to 3 cm across.

Sceletium tortuosum (Mesembryanthemum tortuosum) — Kougoed, Kanna
(Mesembryanthemaceae)

This is not a showy plant, but bears mention as it is the "kougoed", meaning "something to chew", that was chewed by the Namaquas all day long for its pleasant smell and stimulating taste. The leaves are fermented, dried and then chewed. They have been found to contain the narcotic "mesembrine" and the plant was used for medicinal purposes by the Namaquas, who gathered it from the mountains of Namaqualand in October, as noted in van der Stel's Journal in 1895.

This is a spreading low plant, with small, oval, fleshy leaves, placed opposite one another on the stems and becoming skeletonised with age. The whitish flowers are about 4 cm in diameter and are usually produced singly at the ends of the branches. There are several species to be found in Namaqualand and the Karoo.

Stapelia — Stapelia, Carrion Flower, Aasblom
(Asclepiadaceae)

These small succulents always evoke interest when in bloom as they have fascinating star-shaped flowers that emit a smell of bad meat so as to attract flies to pollinate them. They are prized by succulent growers and have been depleted by collectors so that every effort should be made to preserve them in nature.

They have thick finger-like, 4-angled stems that grow upright, forming large clumps, and the flowers emerge one at a time from the base or sides of the stems, remaining open for only a few days until they have been fertilized. They are generally purplish or yellow in colour and often beautifully marked. They are cup-shaped at the base and open into 5 broad petals that taper at the tips. There are 80 species from South Africa and tropical Africa, with several in Namaqualand.

There are several allied genera that occur in Namaqualand, of which *Huernia*, *Caralluma* and *Hoodia* are probably best-known. Those interested should study the detailed descriptions in "The Stapelieae" by White and Sloane.

S. namaquensis. THE NAMAQUALAND STAPELIA.

Although there are many more attractive species, this is notable as it was named for the area and its variable forms have been collected from many different localities. The flat flower measures about 9 cm across and is greenish-yellow, speckled with purplish-brown markings, with a raised rim or *annulus* around the centre. It has often been compared to the variable *S. variegata* from Table Mountain.

S. pulvinata. CUSHIONED STAPELIA.

The gorgeous flower of this species is like a sea-anemone. It is purple-brown, streaked across with yellow lines, while the broad centre, at least 9 cm across, is covered with a cushion of shaggy, soft purple hairs that also fringe each lobe of the flower. These curve outwards and hang down from the central crown. The flowers are solitary near the base of the stems. This species occurs in the Kamiesberg and on the hills around Springbok and Concordia in Namaqualand.

Kinkelbos, (Curly-bush) *Tetragonia fruticosa*
(Tetragoniaceae)

A rough, shrubby plant that grows to a height of about 60 cm, this has red, succulent stems and narrow, fleshy green leaves with rolled edges. It bears numerous tiny, 4-petalled, yellow flowers and red, winged fruits. This species is common in the rocky hills around Springbok, often dotted about in the sandy fields between bright annuals, as well as in the S. W. Cape.

Orange Crassula *Vauanthes dichotoma*
(Crassulaceae)

A small annual succulent of about 20 cm in height, this has tiny fleshy leaves and a flower-head of bright orange flowers, each with 5 slender pointed petals. They may be deep gold and marked with red at the base. Each small flower, about 2 cm across, is borne at the tip of a branched cluster of hair-thin stalks. This little succulent is found growing in the shade of bushes south-west of Kamieskroon and is common in sandy places in the S. W. Cape, flowering in spring and early summer.

Good Books for Reference

Flowering Plants of Africa, illustrated by famous artists like Cythna Letty, with descriptions written by the botanists at the National Herbarium, Pretoria. Forty volumes to date, with a new volume added every 2 years. A Government publication, which may be found in libraries or purchased by private individuals.

Wild Flowers of the Cape of Good Hope, by R. H. Compton and Elsie Garret Rice. Excellent illustrations of many of the same or related species to those that occur in Namaqualand. This is a guide to identification in the field and one of the best handbooks on Cape flora for the amateur.

The Genus Babiana, by G. Joyce Lewis.
Monographs on *Iridaceae* and *Ixia* by G. Joyce Lewis.
Ericas in South Africa, by Baker and Oliver.
The Aloes of South Africa, by G. W. Reynolds.
South African Aloes, by Barbara Jeppe.
Aloes of the South African Veld, by Bornman and Hardy.
The Succulent Euphorbiae, by White, Dyer and Sloane.
The Stapeliae, by White and Sloane.
A Handbook of Succulent Plants, by Jacobsen.
The Genera of the Mesembryanthemaceae, by H. Herre.
The Geology of South Africa, by Alex L. du Toit.
  (Third edition edited by S. H. Haughton).
Gems, Minerals and Rocks in Southern Africa, by J. R. McIver.
The Khoisan Peoples of South Africa, by I. Schapera.
Birds of South Africa by Roberts.

Simon van der Stel's Journal of his Expedition to Namaqualand, 1685-6, edited from the manuscript in the library of Trinity College, Dublin, by Gilbert Waterhouse. (Longmans, Green and Co.)

Catalogue of Pictures in the Africana Museum. Vol. Two. C-D. Compiled by R. F. Kennedy.

Index

Botanical names are in italics and common names in capitals